SECURITY DESIGN CONSULTING

SECURITY DESIGN CONSULTING

The Business of Security System Design

Brian Gouin, PSP, CSC

AMSTERDAM • BOSTON • HEIDELBERG • LONDON
NEW YORK • OXFORD • PARIS • SAN DIEGO
SAN FRANCISCO • SINGAPORE • SYDNEY • TOKYO
Butterworth-Heinemann is an imprint of Elsevier

Acquisitions Editor: Pamela Chester
Assistant Editor: Kelly Weaver
Marketing Manager: Phyllis Cerys
Project Manager: Jay Donahue
Cover Designer: Eric DeCicco

Butterworth-Heinemann is an imprint of Elsevier
30 Corporate Drive, Suite 400, Burlington, MA 01803, USA
Linacre House, Jordan Hill, Oxford OX2 8DP, UK

Library of Congress Cataloging-in-Publication Data
Application submitted

British Library Cataloguing-in-Publication Data
A catalogue record for this book is available from the British Library.

ISBN: 978-0-7506-7688-5

For information on all Butterworth–Heinemann publications
visit our Web site at www.books.elsevier.com

Table of Contents

Foreword

The security industry, in recent years, has witnessed explosive growth in the number of people entering into the specialty of consulting. This frenzied expansion in the security consulting field can be traced back to three pivotal milestones: the September 11, 2001, terrorist attack on the United States; the increase in corporate crimes perpetrated by leaders within public companies; and the blazing pace at which technology is improving our personal security and the security of our businesses. Born from these events, respectively, are *security management consultants, information system consultants,* and *security design consultants.* As with any exigent circumstance in business, there are those who join the ranks of industry professionals with valuable experience and education, and thus something to offer. Invariably, there are also those who jump in simply to capitalize on new market conditions. The security consulting business is no different.

What separates these two distinct types of security consultants is their intrinsic focus. Is the emphasis on boosting the consultants' own bottom line, or is the focus the well-being of their clients? This is not to imply that one cannot be financially successful while serving a client's needs. To the contrary, security consultants who provide their clients with objective and sound advice and demonstrate integrity in their business practices are often rewarded with repeat business and referrals that can generate healthy profits. Ultimately, those security consultants who abide by a strict code of ethics, are independent, and protect their integrity above all last a lifetime despite ever-changing market conditions.

Perhaps more so than other security consultants, security design consultants' independence and integrity are challenged more frequently due to the nature of their specialization. Unlike security management consultants and information system consultants, security design consultants design physical protection systems and therefore must work closely with the integrators, installers, and salespeople whose companies manufacture or distribute physical security products. It is the necessity of working with specific products or particular companies that can create potential conflicts of interest and where the honorable consultant must carefully walk a fine line, carefully balancing his client's needs with security products from companies he is intimately familiar with or from those with which he is less familiar. Brian Gouin coined the phrase *matching product to need* in contrast to *matching need to product.* "MATCH PRODUCT TO NEED not NEED TO PRODUCT." Only after the security design consultant has an understanding of what physical security countermeasures are needed to protect a client's assets can a specific product be matched to that need. While fairly simple to understand, this concept is critical to ensuring that the security design consultant's focus is on the client.

End users, security managers, and directors will appreciate this concept, as most have had the unfortunate experience of a security product salesman, and maybe even a consultant, attempting to sell them a specific product that doesn't fit the actual need of the end user's security program. Unless an end user is intimately familiar with the vast array of security products available in the marketplace today and has kept up with the fast pace of technological change, that user can usually benefit from the services provided by a security design consultant who remains objective and is independent of any security product manufacturer or installation company and whose product choices are based solely on the needs of his clients. As evidence of the industry's need for independent security design consulting, the International Association of Professional Security Consultants (IAPSC), the industry's premier security consulting association, has experienced significant membership growth, particularly among members who specialize in physical security design.

Karim H. Vellani, CPP, CSC
September 2006

Acknowledgments

As with any worthwhile endeavor, the development and writing of this book could not be accomplished without the help and support of many other people. It is impossible to thank everyone who gave input and advice in small but important ways; suffice to say their contribution is very much appreciated. There are, however, many people who deserve special recognition for their help and contributions.

A few years ago I was attending an American Society for Industrial Security (ASIS) conference and browsed the Butterworth bookstore. As I was leaving, a gentleman asked me if I had found everything I needed. Not knowing who he was, I stated that I had not; there was no book dedicated to security design consulting. He asked me if I wanted to write one. It turned out I was speaking to Mark Listewnik, the publisher's editor in charge of security-related books. Thus began the process that culminated with the publishing of this book. I want to thank Mark for his belief in this work and everyone else at Butterworth who put in the effort to make it a reality.

Thanks go to Chuck Sennewald for not only founding the best and most professionally rewarding professional association I know, the

International Association of Professional Security Consultants (IAPSC), but for writing the groundbreaking book *Security Consulting*, whose format was the model for the contents of this book. Many thanks go to the membership of the IAPSC who are so willing to help with any problem, especially to Steve Kaufer and the rest of the faculty of the Successful Security Consulting course, which taught me so much.

I thank Karim Vellani for both his contributions to the contents of this book and for his friendship through it all. Book contributors Sue Meyer and Ralph Witherspoon added great value to their respective subject matters, which I greatly appreciate. John Sottilare from Stanley Works and Robert Martin should also be thanked for their efforts. My father, Don, deserves special gratitude as well, not only for his help with the book, but his sound counsel during the entire course of my business career.

Last and most important I want to thank my wife, Anne. Her love and support made it possible for me to put in the time to write this book while still running a small business and helping her manage a family. It could not have been done without her.

It is my hope that this book will help inspire those thinking of joining the ranks of professional security design consultants and give them the information they need to make the right decisions.

Brian Gouin, PSP, CSC

Introduction

With the growing complexity of technology in the security industry and the relatively poor quality system design work within that industry, there is a growing need for design professionals specifically trained and working in the security field. This book explores the profession of security design consulting and walks through the real-world step-by-step process of what is required to become such a professional and the services the profession provides.

This book gives no specifics about security technologies or the application of those technologies. There are plenty of excellent books available on physical security which cover that subject. Instead, this book is modeled after *Security Consulting* by Chuck Sennewald. Its main function is to serve as a blueprint of the start-up process for someone considering entering the security design consulting profession. In addition, it walks through the details of the services these professionals provide. The book is written with the assumption that physical security countermeasures and the systems encompassing those countermeasures are already known by the reader. Expertise in the physical security field is a requirement for anyone entering the profession of security design consulting.

The function of this book is also to help other industry professionals understand what services security design consultants offer and therefore what impact the consultants have on the security design process and the security industry in general. In addition, the book may also help security directors and others in charge of a company's security function to better understand the services provided by security design consultants and how those services may fit a company's needs during a physical security countermeasure system design and implementation project.

There are chapters in the book that deal with the "how to" for the business end of consulting, such as getting started, marketing, fees, and continuing education. Then there are chapters on the nuts and bolts of the actual work performed by security design consultants such as writing the proposal, performing the assessment, creating the system design, determining the products to specify, determining total system cost, writing the report, and managing the project. The appendices at the end of the book, while lengthy, are examples of actual work products successfully implemented for real projects by security consultants. While they shouldn't be copied verbatim, they can be used as examples and a starting point for a new and perhaps even better work product.

Throughout this book there is an emphasis on quality. Part of the reason security design consultants exist and are needed is the shoddy work product produced in the field for many years by other design "professionals." Most architects, electrical engineers, and security hardware salesmen simply don't have the needed knowledge or experience to design proper physical countermeasure systems. The result is that many systems do not meet the needs of the clients and end users. That poor quality is unacceptable in the security design consulting profession.

Hopefully this book will contribute in some small way to increased professionalism in the design of physical security countermeasure systems within the security industry and the emergence of more qualified and quality people into the field of security design consulting.

1

The Profession of Security Design Consulting

The *American Heritage Dictionary* defines a consultant as "A person who gives expert or professional advice." In the world of professional consulting, the definition should be expanded to say "A person who gets paid to give expert or professional advice." There is an inherent understanding of what a consultant is: a person with knowledge of a specific industry or subject matter who shares that knowledge with the client to help the client with a particular issue. It is also understood that a consultant has some relatively high degree of experience in a particular field doing other work and did not start his career as a consultant. As with other consultants, this is the case with security consultants. The *American Heritage Dictionary* defines security as "Something that gives or assures safety." While the practice of security consulting certainly cannot assure safety, providing an extra measure of safety to people and property more than what is already there is the essence of security consulting.

The security design consultant is also commonly called a *security technical consultant* or *security engineering consultant*. This book will use the term *security design consultant*, which encompasses any of the three terms.

At its most basic, a security design consultant is an expert in physical security countermeasures. Some of these countermeasures include

1. Closed circuit television (CCTV) systems
2. Electronic access control systems
3. Electronic burglar alarm systems
4. Electronic fire alarm systems
5. Gate operator systems
6. Perimeter security systems
7. Lighting
8. Door hardware
9. Locks

Not every security design consultant may be an expert in each of these areas but may specialize in one or many of them. The services offered to clients by a security design consultant are separated into three major categories:

1. *Assessment:* An assessment is an evaluation of the physical security countermeasure needs of the client. An assessment usually begins with a site visit of the client facility, including interviews, observations, and surveys. An assessment involves more than just saying "ABC Company needs to have a CCTV system installed." The assessment includes actual device locations and types, requirements and parameters of all system devices and head-end equipment, and budget estimates. A report with recommendations is usually the outcome of an assessment.
2. *Design:* There are two categories of design services provided by a security design consultant. The first is taking the system requirements and parameters from the assessment phase and making product choice determinations. The results of this aspect of design are usually included in the assessment report. The second is actually writing the design specifications and preparing a set of drawings for the designed physical security countermeasure system. Bidding documents are then attached to the specifications and drawings so the project can go out to bid.
3. *Project Management:* In short, project management services are to make sure the physical security countermeasure system is implemented per the specifications and drawings. Tasks include assistance in locating contractors to bid, running the pre-bid conference, evaluating bids, checking job progress, responding to job inquiries, reviewing shop drawings an as-built drawings,

approving change orders and contractor payments, and overseeing testing.

SECURITY DESIGN CONSULTANTS VERSUS SECURITY MANAGEMENT CONSULTANTS

The most common type of security consultant—the one most thought of as a security consultant with the longest history—is the security management consultant. Unlike the technical expertise regarding physical security countermeasures required by security design consultants, security management consultants have expertise in personnel and policies and procedures. Security management consultants do the following:

1. Evaluate the overall corporate security program.
2. Perform security surveys.
3. Review and write security policies and procedures.
4. Evaluate proprietary and contract guard services.
5. Review and write guard post orders.
6. Train security personnel.
7. Review security budgets.
8. Help hire security staff.

There is some overlap between the two disciplines. Some security management consultants evaluate lighting, locks, and other physical security countermeasures as part of the overall security survey of a facility. They also may recommend the addition or alteration of physical security countermeasures. What they do not do as a rule is determine what the requirements are for those countermeasures, design them, and make sure they are implemented correctly. Conversely, security design consultants may evaluate the performance of guard services as it relates to the use and functions of physical security countermeasures. However, for the most part, they are two different disciplines within the same general professional field and, in fact, work very well in conjunction with each other to offer comprehensive security consulting services to clients. Partnering between the two disciplines on a project is very common and a tremendous asset to clients and the consultants themselves.

QUALIFICATIONS OF A SECURITY DESIGN CONSULTANT

So who becomes a security design consultant? Many, if not most, come from two professions: owning or managing an integration or installation firm (like this author) or being a manufacturer's representative. That is not to say that an equipment salesman, someone working for a manufacturer, or someone with a corporate security background automatically isn't qualified; it just seems to work out that most come from those two fields. The most important point is that the consultant has a solid technical security background and experience in the design and implementation of physical security countermeasures, along with real-world knowledge of the industry dealing with those countermeasures.

A consultant needs to have the credentials not only to appear to be an expert in his field, but to actually be that expert. A consultant will not be a consultant for long if his advice is not sound and his work product is unprofessional, incorrect, or not up to an expected standard of excellence. That will, in turn, shed a bad light on the profession itself as well as the individual consultant. Following are some professional criteria that can be used to determine whether an individual has the qualifications to be a security design consultant:

1. *Work Experience:* Work experience is by far the most important criterion. A consultant must have hands-on experience in his area of expertise. The individual wanting to be a security design consultant must have actually been personally involved in the evaluation, design, installation, and maintenance of the physical security countermeasures on which he wishes to consult. Some of that experience should be in a supervisory position. The consultant must have an understanding of every phase of a physical security countermeasure project process, from assessment to design to procurement to construction and implementation. The expertise must include budgetary analysis, knowledge of available product selections, and overall real-world industry knowledge. It is probable that integration company owners and managers and manufacturers' representatives are the most common people to become security design consultants because they have the necessary work experience for the job. Basically, it is difficult, if not impossible, to consult on a process that has not been personally experienced. It's like taking advice on a surgical procedure from a doctor who hasn't performed one. Why would anyone do that?
2. *Professional Certifications:* In today's world, professional certifications have become a measure of a person's competency in his area of expertise. Whether this measure of competency is real or just a perception is irrelevant; it exists. Therefore, a consultant should have the certifications widely held within his industry as a statement of professional excellence. For most of these types of

certifications, a consultant must meet experience, education, and referral qualifications before applying for the certification. Then he must take and pass a written test to become certified. There is also a recertification process in which certified people must continue to have a strong level of involvement in the industry to remain certified. For security design consultants, the most appropriate certifications include the Physical Security Professional (PSP) and Certified Protection Professional (CPP) certifications administered by the American Society for Industrial Security, International (ASIS) and the Certified Security Consultant (CSC) certification administered by the International Association of Professional Security Consultants (IAPSC). These certifications are described in more detail in Chapter 13.

3. *Education:* Having a college education is nowhere near as important for a security design consultant as for a security management consultant. Such education is simply not expected as much by clients because of the nature of the work. This author does not have a college degree. That is not to say a college education isn't helpful in adding to a consultant's credibility; it certainly will. A college degree does demonstrate a person's work ethic and commitment to completing a project. One college degree that would actually add practical value to a security design consultant's career is an engineering degree (PE). Many security design consultants hold this degree, and in fact, the degree is mandatory for someone wanting to design larger electronic fire systems.

As well as professional criteria, personal criteria are necessary, if not required, to be a successful security design consultant, or any consultant for that matter. The fact is that no matter how professionally qualified a person is, not everyone has what it takes to be a consultant or to be self-employed. Not every security design consultant has to be self-employed; larger firms hire consultants. However, most of the criteria described here apply whether consultants are self-employed or not. In fact, although this book is specifically written with the self-employed small company security design consultant in mind, most of the principles and ideas within these pages apply to all. Some of the personal criteria include

1. *Self-Motivation and Discipline:* A consultant, particularly a self-employed one, cannot be someone who just waits for work to be assigned and then does it, no matter how well it is done. The person has to have the self-motivation to do what it takes to market his services, find and sell the work, and then do an

excellent job. The person needs to be disciplined both in making sure his time is not spent surfing the Internet or playing solitaire instead of working and in meeting deadlines, no matter how late he has to work. Otherwise, the consultant's career will be short lived. Only a small percentage of people have the self-motivation and discipline to be self-employed or even be in charge of their own time if employed; that is why everyone doesn't do it. Take from that small percentage the even smaller percentage of those with the professional security qualifications to be security consultants and then the even smaller percentage of security consultants with the technical expertise to become security design consultants. Not a lot of people are doing this type of work, which is a good thing for those wanting to break into the industry and have what it takes to be successful.

2. *Confidence:* The consultant must have confidence that he knows what he is talking about and will do an excellent job for the client. The client will certainly know otherwise. The consultant doesn't necessarily have to be gregarious, but must be able to relate and talk to people in a manner that makes the client comfortable and confident that this is the right person to help with the problem or project. The best way for a person to have confidence in what he is talking about is to actually know what he is talking about. A consultant can't be an overly shy person or someone who isn't able to speak in front of groups. Communication skills are very important for a consultant. In particular, a self-employed security consultant cannot just do the required work in an office and submit the work for approval. Being able to relate to people on at least a professional level is required to obtain the information necessary for the job to be done well.
3. *Willingness to Travel:* Security consulting requires travel, and a lot of it. Most consultants are not lucky enough to have all their work in a small geographical area around their residence or office. There is an adage that says "Go where the work is," and that is definitely true for security consultants. A percentage of a consultant's time will be spent in airplanes, hotels, and the car. If someone is unwilling or unable to do that, security consulting is the wrong profession for him. As with being self-employed in general, work for a security consulting professional can include long hours.
4. *Ability to Handle Uneven Income:* The fact is that income of a self-employed security design consultant, as with most small businesses, goes up and down, unlike the steady income of a paycheck.

This is even more so for security design consultants than security management consultants because security design work is mostly on a project-by-project basis, and retainers and other recurring income generators are infrequent. There are very busy times when projects are plentiful, and there are more lean times when projects are more infrequent. That is the nature of the business. That is not to say a system cannot be set up in which the business pays the consultant a paycheck regularly, but the business itself may not have a consistent income. This is not inherently a bad thing; the overall income of a security design consultant is very good, but the consultant must understand this uneven income and be able to cope with that reality.

5. *Integrity:* In consulting there are many ways to cut corners, not give the client the best effort possible, or even to flat out steal by overbilling or misusing proprietary information. While someone may get away with this type of behavior in the short term, eventually word gets around, and that person's consulting business will be short lived or at least not optimally successful. Unfortunately, there are these types of people in every profession, and security design consulting is no exception. There is no place for this type of behavior in a successful security consulting business or in the security consulting profession in general. Security consulting, like investment counseling and others, is a profession that has a direct effect on the quality of people's lives and livelihoods. There is a high level of trust between the consultant and client that the work performed and advice being given is in the best interests of the client. To break that trust is shameful behavior.

WHO SHOULD NOT BECOME, OR WHO IS NOT NOW, A SECURITY DESIGN CONSULTANT

One critical word that has been left out of the description of a security design consultant up to this point is *independent*. A security design consultant is an independent resource for the client. Therefore, the consultant should not be affiliated in any way with security guard services or electronics installation, hardware, or software providers and should not accept any finder's fees or commissions with respect to his recommendations. A consultant's counsel should be objective, and recommendations should be presented only on the basis of the client's needs, rather than any outside interests. A security design consultant should not have the inherent conflict of interest that companies do who both consult and install.

A security design consultant is not a security hardware or software salesman. Many salesmen for integration companies have "security consultant" as a title on their business cards. This title sounds better than "sales associate" or even "account executive." However, they are not considered consultants within the professional realm of security consulting because they lack independence. Simply put, it is impossible to give unbiased advice in the best interest of the client, which is the consultant's job, if the consultant is trying to sell a product or, even worse, a particular product. Unfortunately, some clients have difficulty distinguishing between the two, and they don't always get the best advice. Dealing with this perception of what is a security consultant is an ongoing issue all security design consultants must deal with by educating potential clients. While more and more potential clients are beginning to understand this distinction every year, this prevalent misperception is not going away any time soon.

There have been many examples in the past decade of major problems developing from the consulting and/or oversight firm being the same as the service provider—Anderson Consulting being a prominent one. While it is not illegal to "consult" and provide the actual designed system in the security consulting field, it is in the opinion of many unethical, and certainly not the behavior of professional independent security design consultants.

Anyone without a majority of the qualifications as described in this chapter should not become a security design consultant—at least not yet. Consulting is not the forum to learn this trade (that is the trade being consulted on; consulting itself is also a trade). It is necessary to already be an expert in physical security countermeasures before becoming a consultant. That doesn't mean a person has to know everything (who does?), but a degree of expertise is a necessity. If a person tries to consult in a field in which he does not have a reasonable degree of expertise, he is sure to fail miserably. In addition, that person brings down the credibility of the whole trade. That is not to say it is necessary to be an expert consultant before beginning, which is obviously impossible. A consultant, from the first day to the last, will continue to learn about how to be a better consultant. In fact, continuing education regarding consulting, physical security countermeasures, and the security industry itself is critical to the continued success of a security consultant and is detailed in a later chapter.

This book is written with just this theory in mind. There is no education within these pages on the categories or types of physical security countermeasure or the pros and cons of using which countermeasures for what situation. It is expected that this knowledge of the security industry is already known. What this book will help with, this author hopes, is how to take that existing knowledge, add information specific to the services provided by a security design consultant, and translate it into a successful security consulting business.

CONSULTING ETHICS

In this profession, as with any quality profession, its members must follow ethical standards. These codes of conduct are what make the profession one that can be relied on for integrity and honesty. Security consulting is no exception. If the consultant can demonstrate to the client that he follows a stringent code of conduct and ethics, the client will be more apt to trust the consultant from the beginning of the project. That will ultimately help to create a better work product in the end. If the consultant actually follows that code, the client will continue to trust the consultant and the business will flourish. The best available Code of Ethics is published by the International Association of Professional Security Consultants (IAPSC). Here is a portion of that Code of Ethics (see www.iapsc.org):

A. GENERAL

1. Members will view and handle as confidential all information concerning the affairs of the client.
2. Members will not take personal, financial, or any other advantage of inside information gained by virtue of the consulting relationship.
3. Members will inform clients and prospective clients of any special relationship or circumstances that could be considered a conflict of interest.
4. Members will never charge more than a reasonable fee; and, whenever possible, the consultant will agree with the client in advance on the fee or basis for the fee.
5. Members will neither accept nor pay fees or commissions for client referrals.
6. Members will not accept fees, commissions, or other valuable considerations from any individual or organization whose equipment, supplies, or services they might or do recommend in the course of providing professional consulting services.
7. Members will only accept assignments for and render expert opinions on matters they are eminently qualified in and for.

B. PROFESSIONAL

1. Members will strive to advance and protect the standards of the security consulting profession as represented in this code of ethics.

2. Members recognize their responsibility to our profession to share with their colleagues the knowledge, methods, and strategies they find effective in serving their clients.
3. Members will not use or reveal other consultants' proprietary data, procedures, or strategies without permission unless same has been released, as such, for public (or all consultants') use.
4. Members will not accept an assignment for a client while another consultant is serving that client unless assured that any conflict is recognized by and has the consent of the client.
5. Members will not review the work of another consultant who is still engaged with the client, without such consultant's knowledge.
6. Members will strive to avoid any improprieties or the appearance of improprieties.
7. Membership will strive to avoid any improprieties or the appearance of improprieties.
8. Membership in the IAPSC is forfeited upon conviction of any felony or misdemeanor involving moral turpitude.
9. Members will never misrepresent their qualifications, experience, or professional standing to clients or prospective clients.

This is the gold standard for the way a security consultant should act and conduct business. If the words *The security consultant* replace the word *Members* or *Membership* in the preceding code of ethics, it should apply to all security consultants, regardless of membership in the IAPSC.

QUALITY

One of the problematic issues in the security field in general is quality. While there is a lot of very good work in the field, there is also a lot of subquality work at best. This also applies to the assessment, design, and project management services associated with physical security countermeasures, whether done by a security design consultant, integrator, architect, or end user. In some cases the shoddy work product is so common that it is expected and even accepted. For instance:

- When this author was an integrator and a bid request came into the office, the information included within the specifications and the drawings submitted for bid were never the same. NOT ONCE. This phenomenon is actually a running joke in the industry. When the integrator calls to find out which information should be followed, the silence is deafening.

- It is actually common for a company's name other than the client's to be within the design documents because it wasn't edited out.
- Critical information is commonly left out of specifications, such as number of cards required for an access control system or storage capacity requirements for a CCTV DVR.
- Symbols commonly appear on drawings that are not identified in the symbol legend. Likewise, many symbols appear on the legend that are nowhere on the drawings.
- In many cases product choices are not based on the client requirements as determined by an assessment at all, but rather by who is the preferred manufacturer of that designer or, worse, which product needs to come off the shelf.
- Even worse than the preceding is product choices made based on what was used for the last client.
- Having a large number of change orders for a project is commonplace, not because the project has changed so much over time, but because the designer simply didn't take the proper issues into account and the system won't work or can't be properly installed without changes.
- System tests have become, in many cases, a tertiary review to see whether a small percentage of field devices are functional, assuming the rest therefore will be.

To truly professional security design consultants, the preceding situations and many more like them are simply unacceptable. This type of shoddy work in many cases drove consultants crazy when they were in their former profession, and they won't let it happen with their work product. Throughout this book, there will be an emphasis on QUALITY, QUALITY, QUALITY. This is also commonly referred to within the industry and this book as *best practices*. Best practices means exactly what it says: to perform one's duties in the best manner possible. The sad fact is many people want to go through the motions and produce perhaps a workable system, but not the best one possible. What will ultimately separate a professional security design consultant from others who just call themselves security consultants or others who perform security design services is the quality of the work product. Providing the best work product possible should be the goal of every consultant, security consultant, and security design consultant.

ARE YOU READY?

The profession of security consulting in general and security design consulting specifically is great and worthwhile. It is a profession in which helping people to protect themselves and their property is the daily assignment. A security design consultant can also make a very good living. A security professional should strive to obtain all the necessary qualifications and become a security consultant. There is plenty of room in the industry for qualified people who take pride in their work and want a challenge in their careers. This profession certainly isn't easy, but it is rewarding both professionally and personally. Perhaps reading the information within these pages will inspire some to join this most excellent profession of security design consulting.

2

How to Get Started

The decision has been made to become a security design consultant. Now what? It's not possible to just say, "I'm open for business" and get to work. Preparations have to be made, and the business operations need to be established before the business can open. Some of the steps that must be taken before or soon after opening the business are discussed here.

The "Business Plans" section of this chapter was written by Sue Meyer of S. Meyer & Associates, an expert in organizational consulting services.

BUSINESS PLANS

Business Plans: Who Needs Them? What Kind? What Do They Look Like?

Every new business needs a vision and a business plan. Figure 2.1 shows an abstract outlining this vision.

Figure 2.1 Abstract: Every new business needs a vision and a business plan

Who Needs a Plan?

To be successful in business, every consultant needs the following:

- A vision: an idea about which he (and his prospects and customers) is certain, even passionate
- A plan
- Action

Needing talent is a given, and having a rich uncle to fund the business would be nice, but everyone has heard of blockbuster businesses that began with an idea in a garage and just a few dollars.

At this point, for those seriously considering becoming security design consultants and what it will take to be successful, perhaps those ideas haven't been formalized in a way that they can be communicated or progress can be tracked.

Ideas without plans and actions are just a dream. Activities without planning can lead to many starts without results, distraction from goals, and diminished returns and can lower the possibility of success.

What Kind of Plan Is Needed?

In business, there are strategic plans, business plans, and marketing plans, and there are a lot of different layouts and templates to choose from. These plans have been scaled down to two proven, flexible formats that merge the strategic and business planning into a single document. Some firms need or desire a marketing plan, which is usually a separate plan, and is covered in Chapter 3.

The ideal business plan is revisited periodically as a business grows and is flexible enough to take advantage of the past and represent the future.

A Business Plan Helps to Stay On-Track

After someone announces intent to establish a new consultancy, that person may be presented with a variety of actions, opportunities, and decisions very quickly. What should be done first? With whom will the consultant and business be aligned? Given a limited number of resources, which opportunity is the best?

Taking the time up front to create a business plan that aligns with the vision statement will pay off many times over because the business plan serves as a solid organizing and decision-making tool to stay focused and moving toward the ultimate goal.

Typical plans are written with goals, actions, and resource allocations for one to five years (more about the time frame at the end of this section). The business plan doesn't have to be long or complicated, and it doesn't have to take a lot of time to create; in fact, its very nature allows the business owner to break it into "chunks" and work on it in 15- or 20-minute time periods.

One successful small consulting firm has used a simple three-page plan, updated annually for 14 years. Like this firm's plan, a good plan should contain the 10 elements described in the following sections.

1. Provide Background The business plan should include company name, principal name(s), qualifications, business structure (i.e., LLC, corporation), business address, and contact information.

2. Create a Vision Statement Most sole proprietors do not create a full-blown strategic plan, but they do create a *vision statement* that can be communicated to clients, financiers, associates, and family members—all the people and firms that directly or indirectly support the person's success.

The vision statement identifies the purpose of the business, which the customers are, and describes the vision of success at the time victory is declared because goals were accomplished. The vision should be timeless and energizing and can be personal.

Following is a sample vision statement from a consultant who provides driver safety training:

> Our firm will provide professional drivers with training and consultation to make their driving safer, their work and lifestyles more profitable and enjoyable, and their financial futures more secure. When we are successful, our client base will span the nation, our firm will be a recognizable brand in the industry, and our clients will enjoy a 95% driver retention rate.

Here's a sample vision statement from an independent security consultant:

> We offer customized services to create, assess, and test security programs and strategies that withstand the exacting scrutiny of our clients, regulators, inspectors, and auditors.
>
> Clients choose us because they have identified a particular problem and have a goal to achieve; they do not have the resources or knowledge and skills necessary to solve problems they have already identified; and because they require an independent, third-party opinion to confirm a decision or provide alternatives/recommendations and solutions.

3. List Values Values are overriding personal and professional priorities that must be met for the firm to take on an opportunity. Examples are independence, honesty, integrity, professionalism, and other intangibles that set the consultant and the business apart.

Background, vision, and values almost never change. They are the cornerstones of the business and should be included in the business plan.

4. List Products and Services The business plan should also include a list of products and services and the percentage of time and resources allocated to each. In the previous example, the security consultant provided assessments, training, and consulting. In 2005, his time and resources were allocated as Assessments = 20%, Training = 30%, and Consulting = 20%.

Some start-up businesses also identify what they are not in this section. This information serves as an excellent reminder when opportunity knocks or temptation arises that can pull time and attention away from the vision of success.

5. List Clients/Prospects and the Benefits They Receive What is the consultant offering clients from their point of view? This section of the business plan is crucial. It is not a listing of every person or organization (that's part of the marketing plan). Instead, this section provides a high-level description that ties the potential clients to the consultant's lines of business.

In the example, the security consultant has identified clients and prospects this way:

> U.S. commercial, industrial, and private organizations seeking security problem solving, training, and integrated security solutions to meet internal, external, and regulatory demands.

It is critical to think from the client/prospect point of view when writing this section.

One successful sales model is built on the belief there are only six benefits (and thus six ways to market) for any consumer:

- Save time.
- Save money.
- Make time.
- Make money.
- Alleviate fear and/or embarrassment (make more safe/secure).
- Prevent fear and/or embarrassment.

6. List Opportunities Additionally, the business plan should include a simple listing of clients, prospects, even Requests For Proposals (RFPs), in the past, on the radar screen now, and on the horizon for the future.

In the example, the security consultant has had opportunities with corporate financial institutions and government/emergency management agencies in the past year. Assuming these were successful ventures, there's a market for his business in these sectors, and they should be the primary areas for future resources, focus, and prospecting.

7. Identify the Trends The business plan also needs to consider what the clients and prospects are facing. How can the business assist them? The consultant should refer to the six benefits in the preceding section and also consider anything in these categories:

- Legal/regulatory
- Environmental
- Financial/economic including willingness/ability to pay
- Social
- Competition
- Technology
- Political

The action part of the business plan: *the "so what?" section!*

8. Identify What Is Going to Be Done Next is the action section of the business plan. So, what actions will be taken (in the next three months, six months, or one year)? This section needs to be as specific as possible and tie the actions to a date and a goal.

This outline, followed closely, should have helped the consultant drill down (see Figure 2.2).

Good action plans contain the following elements:

- *A simple, clear task:* Develop the standard assessment checklist for the firm
- *Responsibility:* Brian Gouin
- *Time frame:* by September 30
- *Next step:* Use it in two programs by the end of 2007

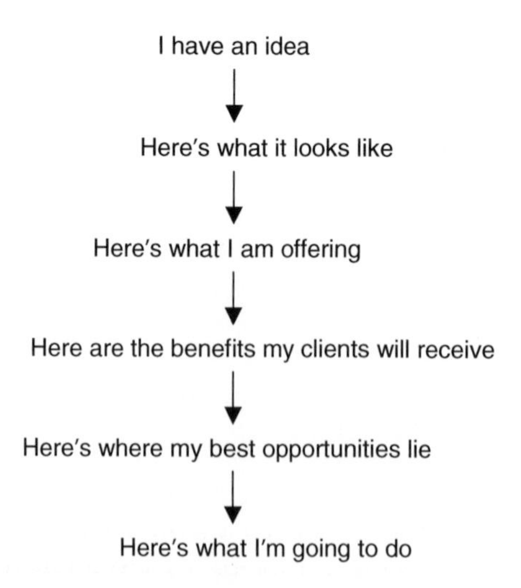

Figure 2.2 Abstract to this point

It is common to have 10, 20, or even 100 actions, and that's great! The key points are to (1) write them down, (2) assign realistic due dates, and (3) track progress.

Outlook Calendar has an excellent task list feature tied to the calendar with reminders and progress tracking.

Not everything has to be done on the 1st, 15th, and 30th day of the month; it's okay to have a task due on the 17th or the 28th. Likewise, most actions don't go from 0% to 100% without some milestones. So actions can be broken into pieces and progress measured along the way to help prioritize and promote a sense of accomplishment.

9. Communicate the Plan The business plan also needs to include who the consultant will tell about this venture, goals, and activities. Sharing the plan is a great way to inspire others to invest in the business and work with the consultant.

The plan also should identify how and how often this information will be communicated. Putting this plan together is an accomplishment in itself, but unless it's used and useful, the planning process and the plan itself are a waste of time and energy.

10. Identify How and How Often the Plan Will Be Reviewed and Updated One of the most gratifying parts of operating a business is taking the time to map out the future, and part of that is reviewing how much progress has been made.

If the business plan contains a lot of immediate actions, it may be used and updated weekly or monthly. If actions are longer term, the review may take place only quarterly or annually. Most business owners take time out of the daily details at least once each year, often in conjunction with a conference or other event, to review their goals and update their actions and plans. How much time? About two thirds of the time it took to create the original plan. Time slips away, so it's helpful to block time or put a reminder on the calendar as the final step in the business planning process.

An Alternative to the 10-Step Plan: The One-Page Business Plan

Much attention has been given recently to a new format called the *one-page business plan*.

The good news is that it works! It contains the same elements as the 10-step plan, but they're stated in bullet format organized on one page.

The other news: Creating the one-page plan requires the same amount of thought and analysis as the traditional plan. However, this is an ambi-

Within the next 3 years, grow CGP into a $3 million national consulting firm specializing in creative leadership development programs for Fortune 1000 companies.

We help companies develop more leaders!

- Increase revenue to 1.8 million in FY 2007.
- Increase gross margin to 54% from 32% by 12/31/07.
- Earn a pretax profit of $450,000 for FY 2007.
- By 12/31/07, establish a client base of at least 10 companies.

- Develop written marketing plan by 2/16/07.
- Trademark "Core Group Process" by 4/4/07.
- Publish 4 quarterly newsletters, send the first to 1500 potential clients on 3/15/07.
- Deliver 5 workshops by 6/30, another 4 in Q3, and 3 in Q4 to net 4 new clients.
- Create high quality company brochure by 7/1/07.
- Create series of four mini-books on new personnel management techniques by 12/31/07.

Courtesy: businessbuilders.bz

Figure 2.3 Sample Plan: Small Consulting Firm

tious and attractive plan, so a sample has been included in Figure 2.3 to encourage prospective security consultants. The format doesn't matter; planning does.

A Final Note Regarding New Businesses: Be Realistic

Any new business takes up to five years to move from "start-up" to established clientele, stability, and profitability. The best-known business gurus and even the IRS recognize this. Why do 97% of new small businesses close their doors within the first three years? Because the owner underestimated the amount of time needed to build a successful business and ran out of optimism, energy, and/or funds to keep going. Therefore, a security design consultant needs to be realistic, even conservative, in planning and share the plan with everyone who will help. After all, isn't it much better to blow past a goal than to underachieve? And won't all of those who helped feel great about being part of a winning proposition?

DEFINING THE SERVICES THAT WILL BE PROVIDED

Before starting a new consulting business, the consultant must define the services that will be provided. Once they have been defined, then the potential clients for those services can be determined as explained in the next chapter on marketing. For a security design consultant, the service choices are basically which physical security countermeasure systems will the consultant provide assessment, design, and project management services for based on his experience and level of expertise. There may be some physical security countermeasure systems for which the consultant has no expertise at all, so those should not be included in the services provided. There may also be particular nuances or niches within any of the three project phases that the individual consultant can offer that others may not. An example of this might be having a particular security clearance that allows for special access to certain facilities to perform these services. Those should be highlighted as well.

BUSINESS NAME

A name must be chosen for the business. There are two main options for naming a consulting business. The first is to use the consultant's name in the name of the business, such as John A. Smith & Associates. The second is a descriptive name based on the type of consulting, such as this author's choice of Strategic Design Services, LLC. If the descriptive name is chosen, the services that will be provided may help determine the name and the name can be a little more creative. There is no right or wrong option for naming the business; there are successful security design consultants using both options. However, the consultant should keep in mind that once the name is chosen and it is registered with the appropriate government agencies, there is no going back. The consultant's advertising, marketing, office supplies, and reputation will be tied to that company name, and changing it later will be very difficult.

BUSINESS ORGANIZATION

The business must be legally formed using some form of business organization. The options include sole proprietorship, S corporation, C corporation, and limited liability company. There are differing laws for different states and pros and cons to each form of business both legally and for tax purposes. An attorney and accountant should be consulted before making this decision. In any case, the business must be legally formed before any work is performed. Many security design consultants do not have any employees and have no intention of having any, ever. However, if the consultant's ultimate goal is to build a large firm with multiple consultants and/or secretarial staff (these do exist as well), he should not choose a sole proprietorship.

FULL OR PART TIME

The goal, of course, is to have a full-time security design consulting business. The question is should the consultant start out part time while he still has a job or should the business be full time from the start? Unless the consultant's existing job has very flexible hours, it would be very difficult to have a security consulting business at the same time as a job. Once the first project is obtained, how will the work get done if there is a job to go to? However, if the job has the appropriate flexibility, it might be ideal to do both right at the beginning, as long as the job does not interfere with the independence of the consultant in any way. It will slow down the business's progression if the full amount of business time cannot be spent marketing the business and preparing for the first client. The consultant's financial position going into the endeavor will dictate some part of this decision as well.

It is a great idea, however, to get everything in place to start the business by doing what is described in this chapter, so the consultant can hit the ground running, while still having the income of a job. Many consultants educated themselves about the profession and prepared themselves for quite some time before actually beginning their businesses. It is, after all, not a decision to be made lightly.

START-UP MONEY

An outlay of money will be necessary to start the business. Some of the items described in this chapter need to be purchased. There are also recurring monthly business expenses such as the phone bill that need to be continually paid before any income is derived from the business. The business's budget is described in Chapter 5 on fees. The consultant needs to make a list of all the start-up costs because they will differ from situation to situation. The other money that needs to be taken into account is the current income of the consultant that won't be there after the consulting business begins. The income for a new consultant will by no means start right away and will take even more time to grow to the point of matching his previous income.

There are varying rules of thumb regarding how much money should be put aside before starting the business on a full-time basis. A good rule is to have all the start-up money plus one year's worth of the consultant's current income before beginning. That will ease any major financial pressure and allow the business to develop and obtain clients in a reasonable amount of time. Many consultants start with less, but wondering where the mortgage payment is coming from is no help when trying to build a business. It is not a good idea to take out a loan for the business start-up money. There will be no income in the beginning to make loan payments, and the loan will be a drag on the business potentially for years. A security design consultant should be able to start and maintain his consulting business without ever having to borrow money.

OFFICE SPACE

There are two options for office space: to put an office inside the home or to rent office space. There are pros and cons to each option. Having an office in the consultant's home certainly saves money and is convenient. Work can be completed at odd hours without having to drive somewhere. However, the consultant must be sure he has the discipline to actually work instead of being distracted or tempted into doing other things because he is at home. If the consultant does not have that discipline, working at home will be a disaster. There also must be a clear distinction between the personal and work space in the home, especially if there are kids at home. It can be a little inconvenient not to have an office in order to meet clients or others, but that situation can be easily worked around. This author works out of his home in an office constructed in the basement.

Renting office space costs money but lends itself to a firm distinction between work and home life. Some people are more comfortable with "going to work" someplace. Some office complexes offer a shared secretary who can answer the consultant's phone or perform other services such as handling any shipping. A more professional appearance might also be achieved to meet with potential or current clients or others. Which option to utilize is a personal preference, and there are professional and successful consultants doing both.

ACCOUNTANTS AND LAWYERS

Accountants and lawyers, especially lawyers, have a bad reputation with some people and probably deservedly so. However, as people say, everyone hates accountants and lawyers until they need one. Accountants and lawyers provide needed professional advice for people in business. The consultant is expecting that other people will pay for and listen to his professional advice, so he should also use the professional advice of others in needed areas. A lawyer will be needed not only to start the business organization but also during the course of the business for contract advice and other issues. An accountant will be necessary for business tax returns and quarterly filings because they are much more complicated than personal returns. Having an accountant file taxes ultimately will save the consultant money. So the consultant should find an attorney and accountant early and use them when necessary through the course of the business.

RECORD KEEPING

The consultant needs to open a business checking account. Some research should be done to find the bank with the lowest monthly fees; however, the bank should be convenient to get to since checks from clients will need to be deposited (with any luck, lots of them). A business credit card or, to start, at least a personal card dedicated to only business expenses is also a good idea, since many items such as travel expenses will require a credit card for payment. Some banks make it very easy to pay bills online. It is very important to keep accurate accounting records for the business. Although an accountant should file all tax returns, it is not necessary and is very expensive for the accountant to keep all the company's financial records. Good pieces of accounting software are available to record financial transactions as they happen, including printing checks. If the consultant works from the home, it may be a good idea to get a post office box for the business mail for privacy purposes, although FedEx is a constant companion of the consultant and those shipments cannot go to a PO box.

It is also important to keep good records that are not financial for both previous and existing clients. Folders should be kept for each client with all the information regarding that client and the work performed. Any reports, specifications, drawings, etc., that were done on the computer for the client should be copied on a disk or CD and included in the client file. It is very common for both past and existing clients to call and request a needed piece of information about the project. It will be much easier if all the information is in one place. Not being able to find something will give a very unprofessional impression to the client.

INSURANCE

Because of the nature of the services they provide, security design consultants require business insurance. Such insurance is not only a good idea for protection, but many clients will require the consultant to have insurance coverage before they will hire him. Liability and Errors and Omissions insurance are the required coverages. Coverage limits average mainly from $1,000,000 to $3,000,000. Only a few companies offer insurance specifically for security consultants, and the policy costs can vary quite a bit, so the consultant must do some due diligence and find a commercial insurance agent who can help.

OFFICE SUPPLIES

The most important office supplies needed are business cards. They don't have to be fancy but should be professional looking to reflect the professionalism of the consultant. All the pertinent contact information should be included on the card, along with short descriptions, even a list of single words, which tell the reader what services the consultant offers. The cards should be printed by a professional printer, never on a personal computer. Along with business cards, letterhead and envelopes will be needed right away. They should have the same professional look as the business cards. All three of these items are relatively inexpensively produced by a professional printer. Other normal business supplies such as paper, pens, stapler, etc., are, of course, necessary. A personal planner of some type is also a business lifesaver. Appointments, to-do lists, and names and phone numbers need to be tracked. Not much is more egregious than missing an appointment with a client.

OFFICE EQUIPMENT

A fair amount of office equipment will be required to perform the functions necessary for a consulting business. Various equipment is described in the following sections.

General Office Equipment

Desks, chairs, tables, bookcases, a drafting table, and filing cabinets are just a few of the normal office items that will be required. They don't have to be fancy or even new to start out. Many office stores sell good used furniture, but it should not look shabby if clients or others will see it.

Computers/Printers

The computer is the lifeblood of a consulting business. Between proposals, reports, online research, and e-mail, the consultant will spend a tremendous amount of time on the computer. A laptop or notebook is a must because traveling is a big part of the business and the consultant will find it necessary to have the computer along on those business trips to do work and check and write e-mail (although a personal digital assistant can also be used for the e-mail). Some consultants also use laptops at the office, so they have only one computer, although most consultants use desktop units at the office because they are easier to use, having a larger monitor screen size and full-size keyboard. The storage size and speed of the computer should be sufficient so it doesn't have to be replaced early on as the business grows. A flash drive or zip drive for backup is also a very good idea.

A color printer is a must so proposals and reports can have a splash of color; also many specifications sheets and other information that will be downloaded off websites have necessary color graphics. The printer should have enough speed so there isn't a long wait for large documents (specifications can be quite long) to print. If the consultant is doing his own CAD work instead of shopping out that service, a printer or plotter capable of producing architectural drawings is required, which can be quite expensive.

Software

Having standard business software such as Microsoft Office is a good starting point. Accounting software and geographic location software are also important. If the consultant will be doing his own CAD work, AutoCAD is the most widely used CAD software in the industry and is very expensive—about $4,000 at the time of this writing—and goes up every year.

Phone

A business phone is a must. It is not a good idea to just have a cell phone for business because of the sketchy service quality. Also, the business phone number should never change, and a cell phone number may need to change if the service provider is changed over time. An answering service can be used to answer calls when the consultant is not in the office, but an answering machine or voice mail with a professional-sounding message is usually fine. The calls can also be forwarded to the consultant's cell phone and switched over to voice mail if no one answers.

Cell Phone

Clearly, a cell phone is also a must. Because the consultant will be traveling part of the time, the cell phone is needed to keep in contact with everyone necessary, check messages, etc.

PDAs

PDAs such as the Blackberry are relatively new devices; they are all the rage and combine some computer, personal planner, e-mail, and cell phone capabilities in one very small device. If the consultant wants to replace some of the previously listed devices with one of these and the functions needed to do business on the road are met, by all means, he should do so.

Copier/Fax

A copier and fax machine are both necessary pieces of office equipment. Some combination units also include scanning capabilities. The fax machine should be a plain paper fax, not the roll paper type, and the copier should have an auto feed feature with reasonable copying speed. A consultant's time is worth too much to be inserting pages constantly into a copier.

REFERENCE MATERIAL

Some reference material needs to be on hand for the consultant to use during the course of his work. These tools include but are not limited to:

- *Code books:* The consultant will on many occasions need to actually read and reference the codes that apply to the application being designed.
- *Dictionary and thesaurus:* These tools help in report writing.
- *Professional directories:* Directories such as *ASIS Dynamics* help in finding sources of information or people who can help with issues that arise.

- *Business help books:* Books on the security industry, business practices, and computer skills are examples of the types of books to have around.

Now that the business is ready for opening, the consultant must market his services to obtain clients and begin work.

3

Marketing

Now that everything is in place and the business is ready to start, the consultant must begin to market his services. The clients are not just going to magically appear; they must be sought out and found. Finding clients takes effort, and this effort will continue throughout the life of the business if the business is going to continue to succeed. One question is, "When is the best time to market?" The answer is all the time, especially when the consultant is busy. This answer may seem counterintuitive, but the consultant won't be busy for long if marketing efforts are not continued during those busy times. The new consultant won't have that problem; there will be plenty of time (100%) for marketing in the beginning. A veteran, successful consultant should be spending 20% of his time marketing for new business to keep coming in and to stay successful. Marketing is especially important for security design consultants because they work mostly on a project-to-project basis and don't have the recurring income of some other consulting disciplines.

The "Marketing Plans" section of this chapter was written by Sue Meyer of S. Meyer & Associates, an expert in organizational consulting services.

MARKETING PLANS

Business planning was introduced in Chapter 2. Marketing planning is a specialized part of planning for the consultant's business. After the decision has been made to be in business, what will be provided to attract clients and solve their problems has been identified specifically, and how to allocate times and resources has been determined, it's important for the consultant to identify how to price, promote, and sell each product or service.

What Is Marketing?

There is often confusion about the distinctions between marketing, advertising, promotions, and selling. The lines blur, and many people resist anything that has to do with the *sell* word. Others say *advertising* is a waste of money. *Promotion* is another word that is tossed around. The following list clarifies how the terms are used and how the pieces fit together:

- *Marketing* is understanding consumer needs and matching those needs. When a company uses its knowledge to design, develop, or modify its offerings to meet customer needs, that's marketing.
- *Advertising* is used to make prospects and clients aware of the consultant's products and services and how to get them. Advertising is typically words and images (also called *messaging*)—in print, on radio, on television, on the Internet, through e-mail and other technology—either purchased or through *press*.
- *Promotions* are tangible goods such as trinkets, giveaways, sponsorships, discounts, donations, or events put on for the target markets.

When a business produces something necessary and convenient for its prospects/clients and is competing with other firms for the business, that is *sales*.

What Should Marketing Plans Include?

While the traditional elements of marketing are product, price, promotion, and place (distribution of products such as retail goods), marketing is competing on bases other than price. For example, Pelco is a closed circuit television (CCTV) system retailer. So is Radio Shack. So is Home Depot. There are target markets and customers for each. How does a consultant reach the right ones, get them to buy, and get them to buy again? Marketing!

If a business plan has already been created, many pieces for a marketing plan are already done. For each product or service, the consultant must define the following:

- Prospects/Clients
- Messages
- Pricing

Many marketing efforts fail because the consultant prematurely decides what marketing messages and tools he is going to use and what to charge without considering how the clients/prospects best receive the message.

Identifying Prospects/Clients

It's important to recognize that the best prospect for new or additional business is *existing clients*. There are four ways to sell to existing clients:

- Offer more of the same product or services to the same recipients.
- Make referrals for the same products/services to other divisions or allied businesses.
- Offer related products or services.
- Bring something "new."

The second best source for new or additional business is *referrals*. A consultant should ask each individual with whom there is a good working relationship for the name of someone else who may need the same or similar services. He also should ask satisfied customers for testimonials.

Beyond that, in the security consultant example in Chapter 2, the prospects/clients were defined as private, commercial, and industrial businesses. Depending on the available time and financial resources, the consultant will further refine these prospects and clients (who needs what is offered) by

- Industry
- Geographic location
- External pressures
- Internal needs
- Willingness and ability to pay

How can a consultant learn some of these things? Some ways are to read local and national news and industry journals; study market information;

listen and learn at conferences, networking opportunities, and meet-and-greets; ask for referrals; read Wall Street filings; and perform other sorts of competitive intelligence. Some businesses engage market research and competitive intelligence services to continuously scan information and identify opportunities.

Targeting Messages

Once a prospective client has been identified, how will the consultant get that client's attention?

First, the consultant needs to identify what he has to offer. The offer may be different for each prospect. The key is to remember the *six benefits*: One successful sales model is built on the belief there are only six benefits (and thus six ways to market) for any consumer:

- Save time.
- Save money.
- Make time.
- Make money.
- Alleviate fear and/or embarrassment (make more safe/secure).
- Prevent fear and/or embarrassment.

How can the consultant turn these benefits into compelling messages for prospective clients? What method will the consultant use (advertising or promotion? print or electronic? trinkets or events?)? The consultant must continue the research because the method varies with every prospect:

- Who in the company does the consultant need to reach? The CEO? Security Director? Risk Manager? Facilities Manager?
- What kinds of organizations does that person belong to?
- What journals does he or she read?
- Where is the firm located, and what types of issues may it be facing?
- Who are the company's vendors? Associates? Parent and subordinate companies?
- What types of research and development is the company sponsoring?
- What can be learned about the company's culture from its website, the press, or research firms such as Hoovers. Does the company sponsor annual events? Does it devote time and money to charitable organizations?

The answer to any of these questions can give the consultant a lead into the firm. How will the consultant introduce himself and his firm? The answer may be different for every lead pursued.

Following are some ways to reach prospects/clients:

- Requests For Proposals (RFPs)
- Advertising
- Direct mail/e-mail/newsletters/press releases
- Community/social events
- Donations or investments
- Person-to-person networking
- Expert testimony to the press (print, radio and TV, trade magazines)
- Blogs, columns, or letters to the editor
- Testimonials/referrals

Following are some ways to achieve free publicity or other promotions:

- Host seminars.
- Send articles of interest, with a business card.
- Give presentations or training; make a speech (advertise in advance, invite the press).
- Link to complementary websites.
- Give an award.
- Make a donation.
- Write a white paper; publish research or findings.
- Market on-line; create a discussion group.

The point is that the consultant must remember to target prospects. Fire with a rifle, not a shotgun.

Pricing

The consultant had some ideas about costs and pricing of products and services to be able to set financial goals and budgets. Formal or informal, every consulting business has a price list or break-even point to be successful. This issue is discussed in detail in Chapter 5.

However, pricing is apart from the rest of marketing because all the effort up to this point has been about getting an opportunity to quote a price or write a proposal. Pricing does not become a factor until an opportunity is on the table.

The consultant needs to resist the temptation to provide a price before having a chance to analyze and customize a solution for a specific client. Any prospect that asks for an estimate and then declines based on budget/price before a formal proposal that weighs the costs and benefits is presented was just a shopper, not a true prospect at this time. This person/organization can be put on the leads list to receive future marketing and promotional materials and occasional check-ins so that the consultant's firm will be top of mind when the prospect is ready. Either way, the consultant should not hesitate to ask for a referral: "Do you know of someone who is in need of my services?" As a result, the leads list will grow quickly.

THE CONSULTANT'S CLIENT

The services that a security design consultant provides determine the clients of that consultant, since the need for the service is a prerequisite to becoming a client. Defining those services was discussed in the preceding chapter. It will be assumed for the purposes of this discussion that the services provided by the consultant are the traditional assessment, design, and project management services for access control, CCTV, and burglary physical security countermeasure systems. The initial potential client pool, therefore, is extremely large and includes the following:

- All existing commercial, industrial, institutional, educational, and governmental establishments and facilities that have a need for these countermeasure systems
- Any new construction of a facility that has a similar need
- Other design professionals who require these services as a part of an overall facility design

Because targeted marketing is more effective than trying to market and appeal to the massive potential client pool just described, the consultant should consider specializing in a small number of industries. Examples of industries include health care, education, airports, casinos, office buildings, chemical plants, financial institutions, manufacturing, retail, warehouses, etc. Prior to becoming one, every consultant had some kind of security experience that probably included dealing with certain kinds of industries more than others. Therefore, the expertise of the consultant is probably more concentrated within those industries. That should be the starting point when choosing potential specialties. If that is not the case, choosing industries in which the consultant feels the services are needed or industries that he has some personal interest in may be appropriate criteria for the choices. The consultant just needs to be sure that if specific expertise is required for the chosen industry, he does, in fact, possess that expertise.

It will be assumed for the purposes of this example that the specialties chosen by the consultant for his business include health care and educational institutions. Now the more manageable pool of potential clients includes the following:

- All existing health care and educational institutions that have a need for these countermeasure systems
- All new construction of similar facilities with the same need
- Other design professionals who require these services as a part of an overall similar facility design

This potential client pool is still very large, but much more manageable to build marketing and advertising campaigns. If the potential client pool becomes too small, the consultant will eventually run out of potential clients. There is no worry of that in this example. The different types of health care and educational institutions can be broken down into categories, such as hospitals and medical buildings, or universities and secondary schools, and specific campaigns can be developed for each category of potential client. Specializing in this manner not only creates more targeted marketing but takes advantage of the assets of the individual consultant. The consultant can become a known expert for this particular scope of services for these particular industries. As more and more projects within these industries are completed, the expertise increases and so does the consultant's reputation as the expert for these industries.

Now, it may be that the new consultant has no idea what to specialize in either because his background includes no particular specialized industry security experience or he just don't know enough to make that decision. That's okay. As the business grows, there will naturally be a tendency for clients to come from certain industries much more than others. Why this happens is unknown, but it always happens that way. This natural growth will then, by default, make the decision as to the consultant's specialties, at least initially. The marketing effort for the consultant should then be a little easier as the target market shrinks.

Just because a consultant has a few specialty industries certainly does not mean that he can't accept and work on a project that is not within those specialties, assuming there are no special qualifications or expertise required for the project that the consultant does not possess. The potential projects that come across the consultant's desk will never be limited to just any defined specialties. Specialties do not and should not limit the opportunities for the consultant but instead enhance them. The consultant's specialties may also change over time because of success or failure with certain types of clients, the acquisition of a large client or large number of clients in a particular industry, or the changing interests of the consultant.

ADVERTISING

There are two types of advertising used to reach clients: those that have the consultant finding the client and those that have the client finding the consultant. It is a good idea to use both types of advertising in an overall marketing campaign.

"Passive" Advertising

The type of advertising in which the client finds the consultant is more of a passive method. Effort has to be put in to develop the advertisement and make it available to the potential client, but after that, it's just a matter of waiting for the phone to ring. However, that does not mean this approach can't be very effective. This type of advertising utilizes the methods described in the following sections.

Website Having a website is a must for a security design consulting firm. Many potential clients find a firm to do business with by searching the web, and many potential clients who find the firm from another source look at the website to find out information about the company and consultant before they decide to call. Websites should be professionally designed because they are a reflection of the professionalism of the company they represent. Additionally, a professional can help word and design the site in a way that helps increase the ranking of the site on search engines. It is very easy to pick out a website that is home-made or designed by students somewhere. Having a site professionally designed may cost up to thousands of dollars, but it is well worth the money. The consultant should do some research before deciding on a designer, because there are literally hundreds of designers out there. He should ask what sites the designer has already designed and go to them to see if he likes the navigation and look of the sites. Figure 3.1 shows the home page of a professionally designed website for a security design consulting firm.

Search engines such as Google and Yahoo! have "pay per click" services through which the consulting firm can pay to have its website come up higher on the list when particular search words are used. This is a very effective method of advertising. A budget is set for a month, and charges are based on the cost paid per click and the number of clicks. The consultant must set the budget realistically for what can be afforded; it is quite easy for this expense to get out of hand.

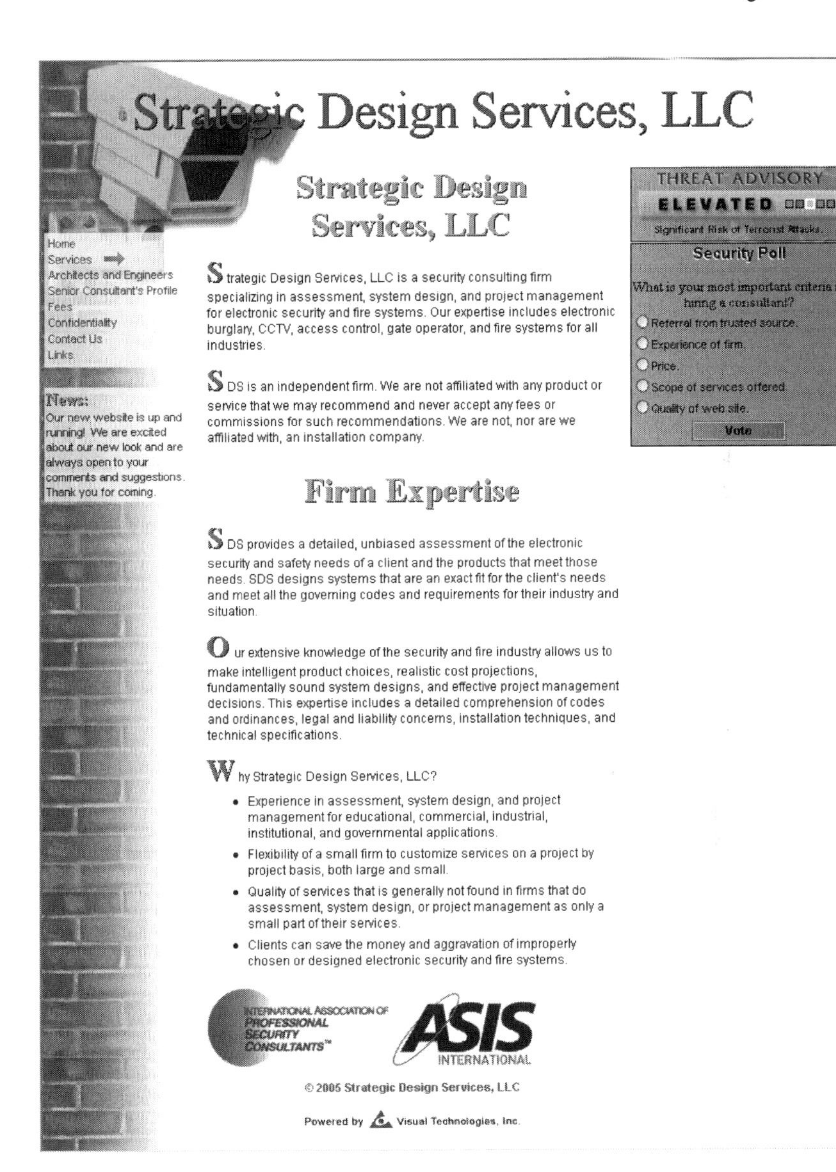

Figure 3.1 Home page of professional website

Print Ads Print ads can be designed and put into trade magazines or industry association periodicals. The smaller the publication, the less expensive the ad will likely be and the more targeted the audience for the ad. That way, the ad can be tailored to the specific service provided for the targeted audience. Advertisements in large magazines or trade papers are most likely a waste of money for a security design consultant.

"Active" Advertising

The type of advertising in which the consultant finds the client is more of an active method. The consultant targets specific individuals or companies to send his marketing materials and then, in many cases, performs some type of follow-up to secure a conversation or appointment. The goal here is to reach a much smaller, more targeted audience than the previous form of advertising. This type of advertising utilizes the methods described in the following sections.

Direct Mail A direct mail advertising campaign is usually directed at individuals or companies within a specific industry in a specific geographical area at any one time. A direct mail campaign can include one or multiple mailings to the potential client within specific time periods. To be most effective, the mailing(s) can be followed up by a phone call to judge the interest of the potential client in the consultant's services. A first mailing, or only if there will be only one, can include a cover letter, business card, and company brochure. Depending on the targeted client, brochures can provide a generic listing of all the services offered by the consultant or a description of the specific services offered for the industry of the potential client. There is nothing wrong with having multiple brochures specific to different industries or services. The important point is to keep them short and small so the potential clients will actually read them and keep them in case of future needs. Figure 3.2 shows an actual generic company brochure, double sided, the size to fit into a normal envelope.

Most of the time, brochures should be professionally printed. They are relatively inexpensive and can be printed on glossy card stock for a professional look. The consultant usually provides the content and format, although many printing companies will help with the formatting for an additional fee. There are, however, also stock brochures available from companies in a tri-fold or other configuration that can be customized for the consultant's individual needs. The format and background graphics are already in place, and the consultant can add his own content and print them on his own computer printer. If this option is chosen, the consultant must ensure that the brochures look professional. Otherwise, the advertising campaign will not be successful.

www.ThreatAnalysis.com

Threat Analysis Group has provided Security Solutions for Business and Government since 1997. We specialize in Security Force Quality Control, Security Program Design & Evaluation, and Risk Analysis. Our experienced security teams include Certified Protection Professionals (CPP), Certified Information Systems Security Professionals (CISSP), Physical Security Professionals (PSP), renowned security & crime prevention experts, & retired federal law enforcement agents.

Threat Analysis Group's distinguished clients include nationally recognized retailers, property management companies, hotel chains, petroleum & chemical companies, security firms, law enforcement agencies, as well as other government agencies. After seven successful years in business, the Small Business Administration has awarded Threat Analysis Group the 8a Small Disadvantaged Business Certification.

SERVICES

QUALITY CONTROL/ASSURANCE

Our experienced Quality Control Inspectors monitor the state of protective forces and ensure contract compliance, training requirements, and proper documentation. Posts are inspected as often as needed, normally on a monthly basis, and reports are submitted weekly or monthly.

SECURITY VULNERABILITY ASSESSMENTS AND RISK ANALYSIS

TAG's security experts perform comprehensive assessments to identify threats and vulnerabilities in your security program. Should the program require modification, our team will assist with the selection of cost-effective measures and implementation. From post order development to turnkey solutions, TAG offers a full range of security services.

INFORMATION TECHNOLOGY SECURITY

Our experienced and certified specialists provide a holistic analysis of your IT security needs. We examine existing infrastructure, provide intelligence about specific vulnerabilities, and offer solutions to keep your data secure, as well as provide awareness training to personnel.

CRIMEANALYSIS™

TAG's early warning system identifies "at risk" properties. The software helps evaluate & select appropriate security measures, and manages vulnerabilities by continuously monitoring system effectiveness. CrimeAnalysis™ accurately determines vulnerability by analyzing the crime history, including temporal characteristics and spatial attributes.

(281) 494-1515
info@ThreatAnalysis.com
www.ThreatAnalysis.com

Figure 3.2 Company Brochure

The second mailing, if there is one, can be a postcard or other such reminder of the brochure that was already sent to keep the consultant's company name in front of the potential client. To be able to manage the direct mailings and follow-ups, the consultant should target between 100 and 300 potential clients at one time during the campaign. Another reason for a small sampling is in case the campaign is not successful, adjustments can be made to mailing content or client lists more quickly.

The purpose of a follow-up phone call after sending the direct mail pieces is for the consultant to determine whether the client has any needs that can be filled with the consultant's services and, if so, to gather the necessary information so a proposal can be prepared. Secondarily, the purpose of the whole campaign is to keep the consultant's company name in front of the potential client so that if and when a need does arise, the client will think of the consultant first and give a call.

Telemarketing Another way of initiating contact with a potential client is through a telemarketing campaign. Here again, professional assistance is probably a good idea. Unless the consultant already has excellent phone skills, likes cold calling, and has the expertise to create an effective script, making the calls himself is not a good idea. In addition, telemarketing takes a tremendous amount of time, and the consultant should be doing paid work instead.

Many telemarketing companies out there can help with scripts and have professional callers who will ensure a professional image for the consultant and the company. As with website development, the consultant needs to do the necessary research, as these services can be expensive, and there are good and bad providers. He also should talk to former or current clients to assess the telemarketing company's professionalism. If the company won't give names of clients, the consultant should look elsewhere.

It is very common in the telemarketing industry to outsource the calling to foreign countries, particularly India. Having American callers is a better idea, for better or worse the results will be better. The goal of a telemarketing campaign should be to set up either a phone or personal appointment between the consultant and client, depending on the geographical location of the consultant and client. The telemarketing firm is not an expert in the security consulting field and should not be discussing project specifics except to gather information for the consultant. It is then up to the consultant during the appointment to determine the client's needs in order to provide a proposal.

Newsletters Newsletters can be sent not only to potential clients but to existing clients as well to keep the consultant's name and company name in front of the client in case the client has other needs that can be filled or knows of someone else who can be referred to the consultant. Newsletters can be sent either by mail or e-mail. They don't have to be very long but should be long enough to be worthwhile reading. Newsletters do not have to be professionally done; most consultants with reasonable writing and computer skills and the right software can easily put together a newsletter and send it out. Figure 3.3 shows a newsletter designed and prepared by a security consultant and sent out via e-mail.

Newsletters should include both new information about the consultant and the company and some industry news or other subject that may be of interest to clients. The idea is to give added value to clients or potential clients so they will actually read the newsletter and think of the consultant when any projects arise. If the consultant chooses to have one, the

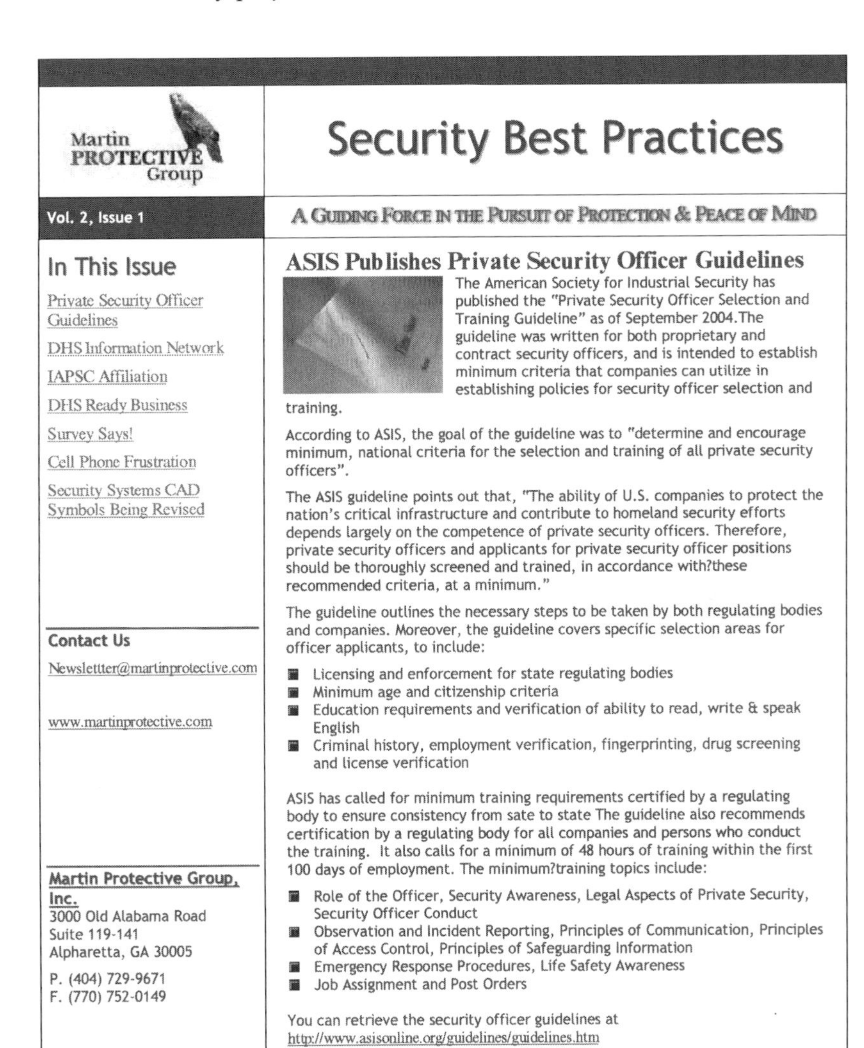

Martin PROTECTIVE Group

Security Best Practices

Vol. 2, Issue 1

A GUIDING FORCE IN THE PURSUIT OF PROTECTION & PEACE OF MIND

In This Issue

Private Security Officer Guidelines

DHS Information Network

IAPSC Affiliation

DHS Ready Business

Survey Says!

Cell Phone Frustration

Security Systems CAD Symbols Being Revised

Contact Us

Newslettter@martinprotective.com

www.martinprotective.com

Martin Protective Group, Inc.
3000 Old Alabama Road
Suite 119-141
Alpharetta, GA 30005

P. (404) 729-9671
F. (770) 752-0149

ASIS Publishes Private Security Officer Guidelines

The American Society for Industrial Security has published the "Private Security Officer Selection and Training Guideline" as of September 2004.The guideline was written for both proprietary and contract security officers, and is intended to establish minimum criteria that companies can utilize in establishing policies for security officer selection and training.

According to ASIS, the goal of the guideline was to "determine and encourage minimum, national criteria for the selection and training of all private security officers".

The ASIS guideline points out that, "The ability of U.S. companies to protect the nation's critical infrastructure and contribute to homeland security efforts depends largely on the competence of private security officers. Therefore, private security officers and applicants for private security officer positions should be thoroughly screened and trained, in accordance with?these recommended criteria, at a minimum."

The guideline outlines the necessary steps to be taken by both regulating bodies and companies. Moreover, the guideline covers specific selection areas for officer applicants, to include:

- Licensing and enforcement for state regulating bodies
- Minimum age and citizenship criteria
- Education requirements and verification of ability to read, write & speak English
- Criminal history, employment verification, fingerprinting, drug screening and license verification

ASIS has called for minimum training requirements certified by a regulating body to ensure consistency from sate to state The guideline also recommends certification by a regulating body for all companies and persons who conduct the training. It also calls for a minimum of 48 hours of training within the first 100 days of employment. The minimum?training topics include:

- Role of the Officer, Security Awareness, Legal Aspects of Private Security, Security Officer Conduct
- Observation and Incident Reporting, Principles of Communication, Principles of Access Control, Principles of Safeguarding Information
- Emergency Response Procedures, Life Safety Awareness
- Job Assignment and Post Orders

You can retrieve the security officer guidelines at http://www.asisonline.org/guidelines/guidelines.htm

Figure 3.3 Company Newsletter

2004 Results of Website Survey

The MartinProtective.com website conducted a voluntary survey for all visitors to the site in 2004, gaining some very insightful information. The survey poll results are as follows:

"What is your greatest security concern in 2004?"

- 41% said Terrorism
- 17% said Computer Security
- 15% said Employee Theft
- 13% said Property Crime
- 10% said Workplace Violence
- 4% said Employee Screening

"For 2004, has your security budget..."

- 57% said it Remained Stable
- 43% said it Increased

"For 2004, has your guarding program..."

- 50% said it Decreased in hours of coverage
- 25% said it Increased in hours of coverage
- 25% said it Remained Stable

"Which Guarding Issues are you most Concerned about?"

- 40% said Security Officer Turnover
- 40% said the responsiveness of the Contract Company's Management
- 20% said Security Officer Training

The Homeland Security Information Network

The Department of Homeland Security has implemented the "Homeland Information Network - Critical Infrastructure" pilot program which you can access at: https://www.seern.gov/forms/enrollrequirements.php The purpose of HSIN-CI is to disseminate real-time information sharing and alerts to those who need to act on that information.

They distribute daily e-mail reports with specific information affecting key critical industries, such as Energy, Banking & Finance, Transportation, IT & Telecommunications, Commercial Businesses and Real Estate industry. In addition, the subscriber receives special reports that contain timely information as it is received and verified through the DHS. Two examples of these reports released in 2004 include "Terrorist Cyber Attacks on Homeland Security" and "How Terrorists Might Exploit a Hurricane".

IAPSC Affiliation

Robert J. Martin CPP, President of Martin Protective Group, Inc has been accepted as a member of the International Association of Professional Security Consultants.

The International Association of Professional Security Consultants, Inc. (IAPSC) is the most respected and widely recognized consulting association in the security industry. Its rigid membership requirements ensure that potential clients may select from the most elite group of professional, ethical and competent security consultants available to them.

The primary purpose of the IAPSC is to establish and maintain the highest set of standards for professionalism and ethical conduct in the industry. The members are independent of affiliation with any product or service they may recommend in the course of an engagement, thus ensuring that the services they render are in the best interests of the client.

More information about the IAPSC can be obtained at http://www.iapsc.org/

DHS Announces "Ready Business" for Terrorism Preparedness

The Department of Homeland Security, has partnered with multiple business and security related organizations in an effort to introduce the "Ready Business" program on its website, http://www.ready.gov. Ready Business is designed to help businesses to prepare in the event of a crisis. The Ready.gov site also helps to inform and prepare families and citizens across the country in the event of an emergency.

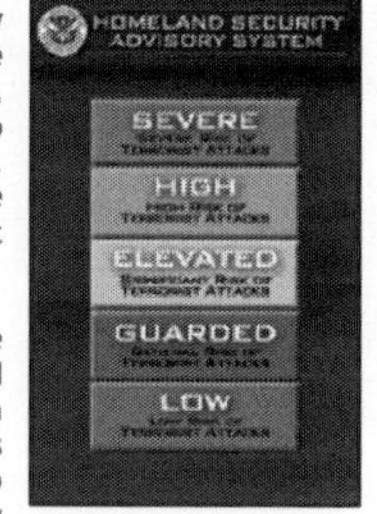

Secretary of Homeland Security Tom Ridge announced that "The terrorist attacks of 9-11 and more recently hurricanes Charley, Frances and Ivan showed that disastrous events can paralyze business operations." "Ready Business was created to help encourage every business to develop an emergency plan, thereby making our nation and our economy more secure."

The Ready Business website (Ready.gov) contains a host of preparedness information to include planning templates and outlines the steps necessary to protect their business.

Figure 3.3 *Continued*

Martin Protective Group, Inc.

Is a widely recognized security management & physical security consulting firm that is passionate about helping businesses and people better protect themselves. We work closely with executives, owners and managers in a variety of businesses, who realize, that now more than ever, security is a critical issue to the very survival of their business.

Our specialized consulting solutions help businesses and security end-users to attain a significant and sustainable return on investment in their physical security programs. Specifically, we consult in the areas of security management, guard force auditing, quality assurance, contract performance metrics, physical security surveys, security awareness programs, policy & procedural refinement and strategic security planning

Cell Phone Frustration

Beginning on January 1, 2005, all cell phone numbers will be made public to telemarketing firms. As of that date, your cell phone may begin "ringing off the hook" with telemarketing calls, however, unlike your home telephone most of you pay for incoming calls. These telemarketers will use up your free minutes and end up costing you money in the long run. To be placed on the national "Do not call list" you need to call 1-888-382-1222 from the cell phone that you wish to have put on the list. You can do the same online at www.donotcall.gov Registering only takes a minute, is in effect for five years, and will possibly save you money and definitely frustration!

Security Systems CAD Symbols Being Revised

Most all drawings for buildings are now made using CAD (Computer Aided Design). Software is used to draw all the walls and doors and windows, as well as all the interior and exterior workings of the building such as electrical devices, plumbing devices, and yes, security devices. These devices are represented by certain symbols that you see on the drawing.

The Security Industry Association (SIA) and International Association of Professional Security Consultants (IAPSC) are once again teaming up to revise and update the SIA / IAPSC Architectural Graphics Standard. For many years, NFPA 170 has been the industry standard for Fire Safety Symbols. However, before 1995 when these two organizations co-sponsored the first Graphics Standard, there was no standard for CAD symbols for the security industry.

With CAD drawings now being commonplace, it is even more critical that there is standardization within the industry. As stated in the purpose of the Graphics Standard: "As many as four different sets of symbols can be seen on a single project: the security manager's set for concepts, the security consultant / engineer for construction documents, the security contractor for shop drawings, and the electrical contractor for installation / as-built drawings."

The Architectural Graphics Standard is a comprehensive set of symbols that are quickly becoming, thankfully, the industry standard. They have been accepted by the American Society for Testing and Materials (ASTM) as a cross reference in their standard of manual symbols. The latest revision and update of this standard should be available in May of 2005.

To download a copy of the Architectural Graphics Standard, please go to SIAOnline.org. Any requests for symbol modifications or additions can be made at Standards@SIAOnline.org.

This Article, Courtesy of Brian Gouin

Brian Gouin, PSP is principle of Strategic Design Services and has over 16 years of experience in the security and fire protection field as a former alarm company owner and current security design consultant. He is a board certified Physical Security Professional (PSP) from the American Society for Industrial Security (ASIS). He is certified in system design from a vast number of manufacturers of security and fire equipment. Brian is a member of ASIS, the National Fire Protection Association (NFPA), and the National Association of Chiefs of Police.

Strategic Design Services, LLC is a consulting firm specializing in assessment, system design, and project management for security and fire protection for all industries. Their design expertise includes electronic fire systems, burglary systems, CCTV systems, access control systems, and gate operator systems. All assessment and design is tailored to the individual client.

Brian can be reached at BDGStrategic@aol.com

Share Your Thoughts

If you have suggestions or Security Best Practices that you would like us to share with others, please email us at newsletter@martinprotective.com

Quote

"In theory, there is no difference between theory and practice, but in practice there is" - Yogi Berra

Figure 3.3 *Continued*

newsletter should be sent out monthly. According to the experts, that is the time period when the company name needs to be put in front of clients again; any longer between newsletters and clients may not remember the consultant's name if a project arises.

COLLABORATIONS/PARTNERING

Partnering with other consulting firms will be mentioned many times within this book. The fact is that while many projects include services that can be provided by a security design consultant, many projects include services that must be provided by others as well. The most common collaboration or partnering is with security management consultants. Many projects include the need for the services of both types of security consultants. Partnering with audio/visual design consultants and IT system consultants is also common. Partnering is an excellent way to pool resources and be able to more competently propose and therefore obtain projects that otherwise might not be obtained.

While a marketing campaign directed toward these other consultants is probably not appropriate, it is necessary for the security design consultant to make these other consultants aware that partnering is welcome and desired. This can be done socially through professional associations or any other forum where there is contact between the two consultants. If the two consultants are working on the same project without being partnered, that is a perfect time to discuss future partnering possibilities. Both sides should keep in mind that partnering is a two-way street. The consultant must be willing to partner with other consultants on projects he encounters if he expects the same courtesy from other consultants with projects they encounter. Also, both sides must be willing to make potential compromises on pricing and other pertinent issues. Any consultant who does not take advantage of potential partnering opportunities is missing a great opportunity to build his business and better learn his trade.

SHOE LEATHER

"Shoe leather" and "pounding the pavement" are both colloquialisms for getting out there and doing whatever it takes. That may mean different things for different people. Some consultants may be very comfortable going into a corporate facility and cold calling the security director. Others may have no problem cold calling over the phone. Maybe going to see every industry contact from the consultant's previous security career is a good idea. Whatever the consultant is good at or has going for him should be exploited, and the consultant needs to get out there and do it. Even if the consultant isn't good at it, he should do it anyway. After all, if he doesn't do everything possible to get clients and make the business a success, who will?

TRACKING SALES EFFORTS

It is important that the results of every advertising and marketing effort are tracked closely. That is the only way to compare the cost of an effort with the results in order to determine if the effort is worth the cost and worth continuing. Not every marketing or advertising effort will be successful. Different efforts are successful for different consultants for a myriad of reasons. The important point is to track the results of the efforts and find the ones that work and do those.

A 1–2% return can be expected from direct mailings. A 3–5% return can be expected from telemarketing. These returns are for appointments, not necessarily sales. The advertising efforts are designed to produce proposals to the potential clients; making the sale is then up to the consultant. For the most part, the total number of sales from these efforts, not just the number of appointments, will determine whether the efforts should be continued. In some cases, not making the sale cannot be contributed to the marketing effort, such as the consultant simply being underbid or underqualified. The returns from the website, search engines, and print ads also can be judged based on number of appointments and number of sales. In some cases, one sale will make the whole effort worthwhile.

The consultant should also track the percentage of projects won versus proposals submitted. It's hard to say what this percentage should be because there are so many variables. If any consultant won 100% of his submitted proposals, his career would be soon shortened by retirement. Between 20% and 50% is a broad range that is probably acceptable. If the consultant has over a 50% success rate, that is outstanding and means he is really targeting his market very well. If the success rate is under 20%, proposals are being submitted to potential clients who have no intention of purchasing the services, the proposals are not of good enough quality, or some other issue needs to be corrected.

PUBLIC SPEAKING

Public speaking is an excellent way for the consultant to promote himself and his services in an indirect way. The subject of most speaking engagements is about some aspect of the security industry or other security-related topic in which the consultant is an expert. These engagements are usually not about what the consultant can provide a potential client. However, by default, members of the audience may have a need for the consultant's services and will remember hearing the speech or presentation and will therefore consider the consultant for a project. Speaking engagements are available at industry conferences (security related or other industries), local civic organizations, and local professional organizations. While it is very competitive to land a speaking engagement at many larger conferences, smaller industry conferences and local organizations are always looking for speakers. The consultant should start small and work up to the larger audiences once the presentations are tried and tested.

If the consultant has no problem with public speaking, this should be an easy way to promote himself and his business. If he does have a fear of or is uncomfortable with public speaking, he needs to overcome the issue by learning how to speak in public. Organizations such as Toastmasters exist to help people with their public speaking skills. Taking one of these courses is an excellent idea, since some type of public speaking, at the very least proposal presentations in front of small groups, is required in the security consulting profession.

WRITING

Writing is an important part of being an industry professional and gives the consultant great exposure, thereby promoting the consultant as an expert in the industry. Published works do not have to be books such as this one, although writing one is a great thing to do, but can be articles in trade magazines, either in the security industry or the industry of the consultant's specialty. Most trade magazines are always looking for articles on different but industry-related subjects, and these articles are a great way for the consultant to begin writing. Also, the consultant can write white papers or information sheets on specific security subject matters to give to existing and potential clients or make them available on his website for free or for sale. In essence, the consultant is self-publishing his work. The important point is for the consultant to have written material attributed to him in order to promote himself as an expert in the industry.

4 The Proposal

Once the business marketing efforts have succeeded and a potential client has been put in contact with the consultant, the consultant must now sell his services to the client to get the job. The word *sell* seems to automatically put a scare in many people. Many consultants have expressed great trepidation at the selling aspect of the business, saying they are good consultants but not good salesmen. Perhaps they associate sales with some pushy used car salesman they have met or the insurance salesman with a million dollar smile and a ten cent brain. In fact, selling for the professional consultant is simply convincing the client that he understands the needs of the client, has the expertise to provide the services needed, and will do so at a reasonable cost. This is done through the verbal communication with the client and through the proposal.

It is true that some people are better than others at communicating with potential clients and convincingly expressing their ability to perform their work. Everything else being equal, those people have an edge over those who do not have these skills. It is also true that communication skills are essential in being a good security consultant, but that doesn't mean someone has to be the "Great Communicator" to be a good consultant or to sell his services to a potential client. The consultant certainly can't have a rude or obnoxious personality, but he doesn't have to be a smooth and completely polished talker either. Communication skills can be learned and improved upon, and every consultant should strive to be better in this area as he grows personally and professionally.

There are also skills that can be acquired about the step-by-step process in the art of selling. Numerous books and other learning materials are available about closing the sale and other such topics, some of which are very good and worth the time to read and make a part of the business practice. Others are not so good, so the consultant must be choosy when implementing these ideas. None of these techniques will be explored in this book; instead, the focus will be on how to find out what the potential client needs and how to present that information in a proposal format. If the reader wants to be educated on techniques to close a higher percentage of the proposals prepared, by all means seek out that information and learn it. Assuming the techniques expressed and advice being given are ethical and worthy of a professional consultant, they can only help.

If the business marketing efforts resulted in the consultant's finding the client, the consultant will more than likely have to devise a scope of services with the help of the client before submitting a proposal. If the business marketing efforts resulted in the client's finding the consultant, the consultant may still have to devise a scope of services with the help of the client, or the client may already have a scope of services in the form of a Request For Proposal (RFP) or some other written form.

INITIAL CLIENT MEETING

If the client does not already have a defined scope of services, the consultant must work with the client to devise one before a proposal can be submitted. Obviously, if the consultant doesn't know the client's needs, it's impossible to demonstrate to the client that those needs can be met. The forum to find out this information is called the *initial client meeting*. This client meeting can take place through several avenues to find out the required information, as discussed in the following sections.

Meeting in Person with the Potential Client

There are several advantages to meeting with a potential client in person. First, face-to-face contact is almost always the best in not only immediately developing a relationship and rapport with the potential client, but also having the most effective give and take in determining the exact client needs. Second, the fact that the potential client will take the time to meet with a consultant may demonstrate that the client is serious about hiring someone to help with a situation; therefore, such a meeting is worthy of the consultant's effort.

There is also a disadvantage to meeting with the potential client in person: such meetings cost time and money. If the potential client is within a short driving distance from the consultant, the cost isn't such a big deal, but if the travel includes airfare and even hotel accommodations, the cost can be significant, not to mention the time spent away from the rest of the business. While these expenses can certainly be included in the cost of any services proposed, if the consultant doesn't get the job, then the expenses must become part of the cost of doing business. If the consultant is not careful, these kinds of expenses can quickly get out of hand.

At a recent International Association of Professional Security Consultants (IAPSC) conference, a management consultant brought up the fact that he charged a potential client for an initial meeting to determine the client's needs. He justified this charge by promising and then delivering something of value to the client during the visit. This revelation caused quite a stir within the group, especially from the security design consultants who would love to be able to do this. In the end, the consensus was, probably correctly, that it would be very difficult for security design consultants to charge for this initial visit based on the nature of the business.

That is not to say that some potential clients shouldn't be met in person no matter what kind of travel is required. In some cases, the consultant will not be considered for the project without a personal meeting; having such a meeting can also show good faith on the consultant's part. It's a matter of risk versus reward and prequalifying the client. Some kind of conversation has to take place between consultant and client to set up the initial meeting, and the meeting is usually requested by the client. The consultant has to determine whether the client is really serious about hiring someone and, even more important, whether there is a reasonable possibility of getting the job based on the consultant's experience. If the potential client is just shopping but likes to meet in person, such a meeting may not be worth the consultant's time and money. The size of the potential job is also a factor. If the meeting is with a large corporation with multiple facilities, traveling halfway across the country has a better upside than if it is a relatively small project. All these factors must be weighed in making a decision whether to meet with a potential client in person. Many times, making this judgment call is difficult but gets easier with experience.

In some cases, the consultant must visit the facility on which the proposal is based to be able to prepare a proposal. Although the size of the facility and other factors can be determined in other ways, some information required for the proposal may be gathered only by seeing the facility. There is no set rule as to when this is necessary; the decision depends on the type and size of the facility and the possible scope of services being offered. If the consultant feels that a site visit is required before a proposal can be prepared for this reason, the same evaluation described previously needs to take place before the expense is incurred.

Using the Telephone

The next best way of gathering the information needed to put together a proposal is to have the initial client meeting over the telephone. This approach is certainly cost effective and, while not being quite as good as a face-to-face meeting, is effective in having the give and take necessary to determine the exact client needs. A rapport can be established on the phone, and the consultant can communicate his level of expertise. This is probably the most common manner of gathering information. Telephone courtesy skills, such as not using the speaker phone, need to be learned and utilized during these business conversations.

Using E-Mail

We live in a business world where e-mail has become a major form of communication. The fact is a consulting business, like many other businesses, cannot survive and succeed without the use of e-mail. It's amazing how getting someone on the phone seems impossible, but that person immediately responds to an e-mail. While e-mail is great for transferring documents, it's not great for interpersonal communications, mostly because it's not personal. It's difficult to express emotion like in person or over the phone, and the give and take of information and ideas can be cumbersome. Even close friends, like business associates, can misinterpret statements made over e-mail.

The constant use of e-mail is not going to go away any time soon, however, so it must be utilized. Many potential clients will make it clear that e-mail is the best way to communicate with them. Particularly clients who found the consultant through a search engine on the Internet will make the initial contact with the consultant through e-mail and wish that some, if not most, correspondence be in that form. But the consultant and potential client cannot communicate by e-mail alone. There must be a blend with either phone conversations or personal meetings to have the give and take necessary to determine the client's needs to be able to put together a proposal. So the consultant should use e-mail but not overuse it.

INFORMATION NEEDED FOR THE PROPOSAL

For a proposal to be detailed enough to demonstrate to the potential client that the consultant is right for the job, the consultant must gather all the necessary information to properly develop a scope of services and write the proposal. No matter in what form the initial client meeting takes place, the same information must be elicited from the potential client. The following questions should be asked:

1. Does the client know what type of physical security countermeasure they are looking for, or are they looking for a recommendation as to the best fit for their needs? In many cases, clients will say such things as "I want an IP-based CCTV system designed." That makes it easy. Other times they have no idea.
2. Is the client interested in assessment, design, and project management services or just one (assessment is the only one that can stand by itself) or some combination? More than likely, the consultant will need to explain what encompasses each of these services. Clients almost always know what they want after the services are explained. In the case of project management services, clients may also pick and choose which of these is desired. Keep in mind that your expertise and ethical standards are a major factor in this decision. There can and should be services that cannot be provided by themselves or without specific other services and still provide a quality and correct work product. Consultants should never provide less than the needed services for the project just because that's what the client wants. There may even come a case where the consultant needs to walk away from the project for that reason.
3. Is there a budget for the consulting services? There may be one for which it would behoove the consultant to tailor the extent of the scope of services to fit into the budget. Many times, there isn't one and estimating one is a crap shoot.
4. Is there a budget for the installation of the desired physical security countermeasures? This information may help the consultant tailor the scope to a system that will fit into that budget. If there is no way the requested system can fit into the budget, the consultant's expertise may be demonstrated right away by pointing out that fact. Also, mentioning the budget in the proposal is a nice touch and shows the consultant is paying attention to details.

5. Is there a schedule for the project? This information allows the consultant to demonstrate that the schedule will be met and also may affect the cost of travel expenses if the time frame is tight.
6. Are there drawings available of the facility, and if so, are they in AutoCAD? If a design specifications service will be provided, this has a great effect on the cost of those services.
7. Is there anything particular to the client or project that should be known before the consultant submits a proposal? This and other similar all-inclusive questions should be asked to make sure all the necessary information has been gathered.

Once all these questions have been asked and answered, the information is then used to develop a scope of services for the project so that the consultant can write a proposal for the client. Proposals take a lot of time to put together, so the consultant must be sure to gather all the information to compose a quality one.

PROPOSAL ELEMENTS

The exact format of proposals is as varied as the consultants. This includes the naming of sections, the order of those sections, and the way in which the information is organized within the sections. Everyone has an individual style, likes, and dislikes. One format may be equally as effective in getting the correct message across to the potential client as the other. The important point is not the exact format of the proposal but making sure all the elements are present to demonstrate the consultant's ability and expertise to do the work and sell the proposed services.

Appendix A shows two actual proposals submitted to potential clients. The first is a proposal submitted from a security design consulting firm. The second is from a security management consulting firm in which the security design consultant partnered with the management consultant on the project. Notice there are similarities and differences between the two proposals, yet both accomplish the goal. Elements within a proposal are described in the following sections.

Cover Letter

Every proposal should include a cover letter thanking the potential client for the opportunity to bid on the project and explaining any particular aspects of the proposal that should be examined closely. This cover letter can actually be part of the proposal itself or a separate cover sent with the proposal. In the two examples in Appendix A, one includes a cover letter and one does not.

Summary

A proposal should begin with some kind of summary regarding the project goals and what general services the consultant will be providing. This may be called *summary, executive summary, project objectives, project scope,* or other equally good terms. The idea here is for others to be able to read the first section of the proposal and know what they will be reading about. The summary should not be so detailed that the rest of the proposal won't be read, but it should have enough detail that the readers understand the parameters of the project.

Company Information

Some information and sales language should be included in the proposal describing the consulting company, its philosophy, and what the approach will be to the project. This is where the consultant should describe all the attributes of his company and explain why the company will do an excellent job with the proposed project. This can be one section or broken down into multiple sections that may be called *project approach, company expertise, company background,* or other such terms. It's okay to embellish a little, but this information should be realistic and never promise something that can't be delivered.

Consultant Information

The proposal also should include a biography of sorts of the consultant or consultants working on the project or the résumé of the consultant. This section of the proposal is mostly a fact-based representation of the qualifications of the consultant within the industry and in relation to the specific project. This information may be altered slightly from project to project, but, of course, the basic facts will remain the same. Back to ethics, the consultant should by no means lie about qualifications because doing so will come back to haunt.

Scope of Services

The scope of services section is the nuts and bolts of the proposal, the actual description of what services the consultant is going to provide and how they will be provided. It is a step-by-step explanation, usually in chronological order, of the project process. Most of the time, the project is broken down into phases with a brief but detailed explanation of the work items or tasks that will be performed within each element of the phase. The explanation should be detailed enough that potential clients can understand exactly what they are getting for their money without having to ask a lot of questions. This section is usually called *scope of services* or *methodology*, or both are used with *scope of services* being an overview and *methodology* being the details. If this section is not clear or shows a lack of understanding of the project, the chances of being awarded the project are slim. The consultant must pay attention to the details.

Engagement Considerations

Some of the items that may be included in the engagement considerations section are responsibilities of the client, timeline, and project cost. Many times, some or all of these items are separated into their own sections, particularly project cost. Again, the way the information is presented is a personal preference decision. Project cost should never be the centerpiece of the proposal; instead, what the benefits are to the client by hiring the consultant should be first and foremost. Project cost will be discussed in Chapter 5 on fees and billing.

References

One issue regarding proposals is whether to voluntarily supply references. There are two schools of thought about this, and there is no right or wrong answer. Some consultants have strict confidentiality policies that are explained with a proposal, and therefore no referrals are given. Others feel referrals are an important element in demonstrating that they have the requisite experience for the project. Of course, for new consultants, there is the timeless issue of which comes first: the chicken or the egg. How does a consultant gain experience without being awarded projects, and how is a consultant awarded projects without any experience? This question has no good answer, but new consultants can take comfort that every other new consultant had the exact same problem, and all won their first project and gained experience. Some clients will insist that they receive references regardless of the consultant's policy. A decision then has to be made to provide them or walk away from the project. If references are given, the consultant must make sure the projects are as close to the scope of services described in the proposal as possible or the clients are of the same type, such as a municipality.

Ancillary Information

A final section is an all-inclusive category for other elements the consultant feels are a helpful part of the proposal. At the end of the first example in Appendix A is a Code of Ethics that falls into this category. The consultant must be careful not to put too much fluff in the proposal because it may begin to look as though he is trying to hide the real information or lack thereof with a lot of ancillary information. This section should just enhance the rest of the proposal.

RESPONDING TO AN RFP

If the potential client has a defined scope of services, it may be expressed in the form of a Request For Proposal (RFP). An RFP is a written document transmitted from the potential client to the consultant. In this case there is little to no amount of work that needs to be done to understand the needs of the client; these needs should be clearly expressed in the RFP. Some questions may have to be asked to clarify some of the requests or information within an RFP—as with anything else, some are written better than others—but for the most part all the necessary information to put together a proposal is included within the RFP. When an RFP is written, the project is almost always part of a competitive bidding process. Few potential clients put the time and effort into writing an RFP if they plan on having only one or two companies bid the project. It does happen, but it is not common.

RFPs are most common with government projects, from municipal to state and federal projects. They are put out to public bid and advertised in order to receive the most number of qualified bids as possible. Once responses are received and reviewed, in many, if not most, cases a certain number of bidders are asked to be interviewed in front of a client panel. Those presentations are then graded (the details of the grading process are a cottage industry within itself), and the project is awarded to the highest graded consultant. A consultant needs to be prepared for the possibility of this travel when responding to a municipal RFP. New consultants should be choosy as to which of these types of RFPs they respond to, as there will more than likely be many responders, and a consultant with little experience is less likely to be awarded the contract. That is not to say responding to any RFP is not a learning experience that will be helpful for future bidding.

RFPs are not as prevalent for private clients but are still used quite a bit, especially for larger corporate clients. They have the same qualities as government RFPs except they probably won't have the same hiring regulations or be advertised in the same manner—basically, the normal differences between any government and private work.

Putting together a proposal in response to an RFP is quite different from a proposal for which the consultant gathered the necessary information needed. In most cases the RFP will spell out exactly what information is required in the proposal response. The consultant must include all the requested information; otherwise, the proposal may be deemed unresponsive and won't be considered for award. Including other information not requested may be frowned upon and in some cases is disallowed within the language of the RFP. Also, in some cases the format of the proposal response is clearly defined, including the number of maximum pages allowed for each section of the proposal response. References are almost always requested and are mandatory for an RFP response, especially a government one. The consultant should not deviate from the requested proposal format in an RFP; otherwise, being awarded the project is unlikely. The thinking is if the consultant can't do what the RFP asks, how good is the work product going to be?

There is no set format or length to RFPs. They are as varied as the authors themselves. Some are very short, a page or two, but still get across what is needed in the response. Others are quite long; 30 pages is not uncommon. There is also no standard set of requested information. As a general rule, the longer the RFP, the more specific and detailed the information requested and more stringent the requirements of the proposal response. Shorter RFPs typically allow for more flexibility in the response, so they can be more like the consultant's normal format described previously.

RESPONDING TO AN RFQ

Another form of requesting information from a consultant is a Request For Qualifications (RFQ). An RFQ is similar to an RFP except that no scope of services is provided to give a response. Instead, an RFQ is the first step in a two-step process in ultimately obtaining a complete proposal. The RFQ asks only for the qualifications of the consultant, although the format of that request is as varied as with an RFP. Those qualifications are then reviewed by the client, and then a few of the highest qualified submissions are asked to participate in the second step by being given the scope of services so a complete proposal can be submitted. An interview process similar to the one for an RFP can be done either before or after the scope of services is defined. In fact, in some cases the reason there is an RFQ instead of an RFP is that the client wants to work with the consultant to define that scope of services. Inexperienced consultants should shy away from responding to an RFQ in the beginning because the only thing being evaluated is experience and qualifications, including references. An understanding of the scope is irrelevant because there is no scope. Even a lot of experienced consultants do not like responding to an RFQ because of their very competitive nature.

FOLLOW-UP

Now that the proposal is in the hands of the potential client, the waiting game begins. This is definitely the most frustrating time for a consultant, waiting to see if the proposal is accepted and the actual work can begin. This process can be maddening. Some RFPs give an exact date for an award decision, but many do not. Certainly, a proposal without an RFP has no definitive response date. The consultant now has to decide how to properly follow up with the potential client.

One thing that the consultant should avoid is doing nothing. The consultant does need to demonstrate to the potential client through some kind of follow-up that the job is important and that he is very interested in performing the work for the client. The first phone call or e-mail, depending on the established preferred form of communication, should be to ask if the client received the proposal and had any questions. If there are any questions or clarifications requested, the consultant can address those issues before a decision is made. It is important during that first follow-up to ask about the schedule and timing of a decision. The question can be phrased in such a way as to express a desire not to bug the client but to know when a good time for the next follow-up is. The answer to that question will then dictate when the next phone call or e-mail should take place. Regular follow-ups should take place until a decision on awarding the project is made. Following up is a balancing act of not being a pest to the client and keeping the consultant's name in front of the client when the decision-making process takes place. This is where learned "closing the sale" techniques might come in handy.

CONTRACTS

After the extensive proposal process has been gone through and the consulting company has been awarded a project, the consultant must come up with an agreement with the client to provide those services. This is in the form of some kind of contract, often also called a *service agreement*. Regarding contracts, the consultant should consult an attorney. There may be different laws for different states regarding contracts, and an attorney will be able to help with these issues. If anything written in this section contradicts what the attorney says, the reader should ignore what is written here. Remember, use the experts around you.

If the client uses a purchase order (PO) system, that PO is typically the contract and no other paperwork is necessary. A purchase order is a tool used by purchasing departments to verify that monies are set aside for payment. The PO will refer back to the consultant's proposal in some way, and the fact that the PO is issued is agreement by the client. POs will be discussed in more detail in the next chapter. The consultant may want an additional contract signed for liability purposes, but most of the time, this is not done. If a PO system is not used, the client may have a standard contract that is used for all contracting purposes. In negotiating with the client, the consultant might need to use the client's contract form. That contract may be generic and have many sections not applicable at all to this particular project or to a consultant acting as the contractor. Most of the time these discrepancies are manageable and can be worked out. Again, the consultant should consult an attorney.

If the client does not have a standard contract or does but is willing to use the consultant's contract, the consultant should have a standard contract or service agreement available to be quickly adapted to the project and signed by both parties. Appendix B shows one such standard service agreement. Once the contract is signed by both parties, the project can begin.

5

Fees and Billing

The amount of the fee a consultant should charge clients for services provided is obviously a very important aspect of business. If the consultant does not charge enough, in no time he will be out of business because not enough money is being brought into the business. If too much is charged, clients may be unwilling to hire the consultant, and the business will suffer as well. In other words, what are the consultant and his expertise worth in the marketplace? The answer to that question may be different for the consultant versus the client and even from client to client. There are three main elements in determining what to charge the client for services:

1. What hourly fee should be charged for services?
2. How long will the project take?
3. What about expenses?

HOURLY FEES

The first step in determining the hourly fee the consultant should be charging to clients is to go through an exercise using a fee formula taking into account business expenses, desired income, and profit. This formula and associated analysis comes from Ralph Witherspoon, CPP, CSC, of Witherspoon Security Consulting. It's so simple and obvious that it's ingenious and probably hardly ever done. The first time this author saw this formula was at a Successful Security Consulting course sponsored by the International Association of Professional Security Consultants (IAPSC). This exercise is a great starting step for new security consultants and also should be repeated every year or so by veteran consultants to make sure the fees they are charging are keeping up with the times. The formula is as follows:

$$\text{Fee} = \frac{\text{Salary} + \text{Overhead} + \text{Profit}}{\text{Time}}$$

Salary

The first part of the formula is salary. In other words, how much money does the consultant want to make? The temptation when trying to determine what fee to charge is to arbitrarily come up with an hourly fee, and whatever is left over after everything else is paid is the consultant's salary. Instead, the consultant should start with the desired salary and base the hourly fee from that. For the example used in this section, a salary of $100,000 will be used.

Overhead

Overhead is all the expenses of the business, all the day-to-day costs. Every expense must be accounted for; otherwise, the desired salary will be reduced to pay for the expense. These expenses will not be billed to the client, unlike travel expenses, but they are actually costs for the business to operate. Following is an example of all the monthly expenses for a security consulting business:

MONTHLY OVERHEAD	
Office	
Office Rent	$700
Utilities	$150
Other_______________	0
Clerical	
Secretary	0
Communications	
Telephone	$85
Cell Phone	$60

Internet Connection	$20
Postage	$60
Overnight Mail	$40
Other_______________	0
Marketing & Advertising	
Printing (Stationery, brochures, etc.)	$50
Search Engine	$300
General Advertising	$100
Travel (not billed to client)	$500
Other_______________	0
Professional Services	
Accountant	$50
Attorney	$40
Payroll Services	$80
Other_______________	0
Dues	
IAPSC	$46
ASIS	$13
NFPA	$11
Professional Development	
Conferences	$350
Training seminars	$50
Other_______________	0
Business License/Registration	
State Business Registration	$13
Certification Registration	$10
Other_______________	0
Taxes	
FICA	$612
State	$40
Other_______________	0
Insurance	
Liability & E&O	$216
Other_______________	0
Office Supplies	
Copying/Printing	$20
Supplies	$30
Other_______________	0
Equipment	
Computer and Software	$50
Copier, Fax	$20
Other_______________	0

Personal Benefits	
Health Insurance	$550
Vacation Pay	$320
Other_______________	0
Total Monthly Overhead	$4,586

This worksheet should be customized to the specific overhead expenses of the consultant; categories may be added or even in some cases subtracted from what is shown. However, even if some expenses are not currently incurred, it may be a good idea to still account for them in the formula. For instance, if the consultant works out of his home instead of from an office, there is no direct office rent expense; however, there are indirect expenses from having a home office, so it should still be included as a line item. To calculate the annual overhead expenses, multiply by 12 months, which makes them $55,032.

Profit

The company needs to make a profit after paying the consultant's salary. For new business owners, this may be a unique concept, but there is a difference between what the consultant makes and what the business makes, even if it is a one-person business. This profit is used for the long-term growth of the business and as a rainy day fund. This amount is a percentage above and beyond the salary and overhead, usually anywhere from 5–20%. For this example, it will be 10%.

Time

A consultant cannot bill 100% of his time. It is just not possible. Not only doesn't everyone actually perform billable hours 100% of the time, but the reality is other things need to be done in business besides billable hours. The following chart shows an example of unavailable days, thereby coming up with the number of potential billable days:

	Days Per Year	**365**
Weekends (52)	–104	261
Holidays	–10	251
Vacation	–14	237
Sick Days	–5	232
Professional Development	–15	217
Marketing	–36	181
Administration	–24	157

Using this example, there are 157 available billing days. Prevailing wisdom in the industry says billing 50% of a consultant's time is very good. The preceding example is a little less than that but can be adjusted to the individual.

The formula can be used to determine the necessary hourly fee:

$100,000 Salary ÷ 157 Days = $637 per day
$55,032 Overhead ÷ 157 Days = $351 per day
$15,503 Profit (10%) ÷ 157 Days = $99 per day
Total = $1,087 per day, or $136 per hour.

MORE ABOUT FEES

The preceding exercise is a great start in determining the appropriate hourly fee, but other factors must also be taken into account. The "standard industry rate" is something that must be known and considered, especially in a competitive bidding situation. At the time of this publication, the standard industry rate for security design consultants is between $100 and $150 per hour for private clients and between $80 and $130 for government clients. Where an individual consultant should be within this range depends on the experience and qualifications of the consultant, the particular scope of the project, the geographical area of the consultant and client, the particular type of client, and the results from the preceding exercise. There is nothing wrong with changing the hourly fee from project to project based on the totality of the circumstances, as long as the consultant is not unethically gouging a client because the project is sole source or for some other reason. It is a delicate balancing act between pricing the project so the most legitimate amount of money is made and pricing the job too high so it is lost and no money is made at all. Being successful in this delicate balancing act is a constant learning experience for every consultant, no matter how experienced. The hope is that every consultant gets better with it over time.

When the security design consultant partners with security management consultants or others, flexibility will be required in the fees charged. This is certainly the case for the consultant being subcontracted and may be the case for the prime consultant. The degree of flexibility will depend on the consultant's particular situation, schedule, and comfort level. Some consultants are not flexible at all with their fees when potentially being hired by other consultants, and invariably they do not get the work. This really makes no sense unless the consultant is just too busy to care. Partnering with other consultants is almost always a great learning experience and opens the doors for potentially other work that may be billed at normal rates.

In some government bidding situations, mostly on the federal level, the hourly fee is dictated by the government ahead of time and cannot be altered. Therefore, the bidding is based on the amount of time the consultant feels the project will take and the qualifications and experience of the consultant. New security design consultants should avoid this type of bidding process until they are experienced.

One issue that should be pointed out is that the desired salary of the consultant under this formula is achieved only if the consultant actually bills the total number of hours used in the formula. Obviously, for new consultants, this probably will not be the case, so the salary will be less than desired. A major goal, if not the most important goal, of a new consultant is to build the business to a point where the desired number of billing hours is achieved, and therefore the salary goal is also achieved. The fact is that even experienced consultants do not make this goal every year, and it is a constant goal of the consultant to do so. The good news is there are also years in which the number of actual billable hours exceeds expectations, and therefore so does the salary.

Travel Time

There are different schools of thought on the billing of travel time. Different consultants have different policies, and there is certainly no right or wrong policy. Some consultants bill full travel time, portal to portal, including sitting in the airport, etc. The thinking is that this is time that otherwise could be billed to other projects, so it should be billed to the client. Others may use different combinations, such as billing only one-way travel or billing full travel time but only at half price. Still others do not bill travel time at all (travel expenses are billed, of course), either to get a competitive edge or because they feel such billing is unwarranted. As with the hourly fee, there is nothing wrong with this policy changing from client to client based on the geographical location of the client, method of travel (air travel versus driving, etc.), type of client, and so on.

FIXED JOB PRICE VERSUS HOURLY

Security design consulting projects are usually billed in one of two ways: by the hour plus expenses or with a firm fixed up-front price. Billing by the hour is just as it says: the client is told what the hourly fee is, what the travel time policy is, and how expenses are accounted for. In addition, some estimation of the time the project will take is provided, so the total bill to the client is not a mystery. In many cases the consultant will provide a "not to exceed" price so the client knows the maximum that will be charged for the project. The consultant then can bill only the maximum of that amount, regardless of the amount of time spent and expenses incurred. It is not common for security design consultants to bill in this manner.

Most of the time security design consulting projects will be billed using firm fixed up-front prices. The types of projects performed by security design consultants lend themselves to this type of billing because the scope of services is rarely open ended so that an up-front cost can't be determined. In addition, most clients like and appreciate knowing exactly what a consulting project will cost before it begins. If the consulting project cost must be budgeted by the client, it is almost a must to know the exact cost up front. Of course, there is no margin of error for the consultant in this type of billing; what was quoted for the price is what must be billed, regardless of time spent or changes in expenses incurred. There could also be a combination of billing types to a single client such that the main project is billed as a firm fixed up-front price, but then if small additional services are requested, rather than try to estimate the time for those services, the consultant can agree with the client to bill at a specific hourly rate. This approach is actually quite common.

Even if a project is billed on a fixed-price basis, an hourly fee amount still must be determined. That figure is then used along with the estimated amount of time to complete the project to determine the firm fixed up-front cost for the project.

HOW LONG WILL THE PROJECT TAKE?

The next step in determining what to charge the client for a project is estimating the amount of time the project will take. Notice the preceding sentence says *estimating* and not a more definitive term. A consultant makes the best educated guess possible as to how much time a project will take. As the consultant becomes more experienced, the educated guesses become better because of the knowledge of the amount of time similar projects took in the past. What is critical in this process is keeping track of the time that is actually spent on the work for a project. While this may seem like an obvious thing to do, it is surprising how many consultants do not keep an accurate track. Of course, if the project is being billed on a straight hourly basis, the consultant is forced to keep an accurate track; to just guess at the end would be unethical. However, if the job is being billed on a total project basis as described previously, some consultants don't think tracking time matters. Failing to keep track of time leads to making an inaccurate estimation time and time again and potentially costing the consultant money or work. This tracking can be done on any number of pieces of software designed just for that purpose or can be done manually. The important point is that it is done and then is referred to when bidding another project of similar size or scope. The information will be of no good unless it is used.

One effective way of estimating project time is to break the project down into the different tasks as described in the Scope of Services or Methodology sections of the proposal. Table 5.1 shows a sample blank Project Pricing Form. This type of form can be custom designed in Microsoft Excel or other similar software. Mathematical formulas can be entered into specific fields so the totals are automatically calculated. Table 5.2 shows a form partially completed. Individual tasks from the Methodology section of a proposal are listed and assigned an amount of time for the task and for travel if applicable. The total hours and associated dollar amounts are automatically calculated and totaled in the appropriate fields.

EXPENSES

The expenses incurred in the course of completing a project need to be billed to the client. Most of the time expenses are estimated and billed based on actual expenses with a few exceptions such as per diems, as described in the following paragraphs. Expenses may include any of the items described in the following sections.

Subcontracting

Are any subcontractors being used on the project, for instance, for AutoCAD, drawing scanning, or other specialty services? Are any subconsultants being used or partnered with who have given a firm price for their services? All these types of expenses can be accounted for in this manner.

Travel

Some kinds of travel expenses are associated with almost every project. They include airfare, hotel, rental cars, taxis, tolls, parking, gratuities, and meals. While airfare and the associated parking are figured on an actual cost basis, most of the time the rest of the travel expenses listed here are dealt with on a per diem basis, which means a fixed daily fee is charged to include all of these expenses. This makes the expenses much easier to track for billing purposes (especially when dealing with cash for gratuities and the like) and for estimating purposes. In some cases the client's company policy does not allow for a per diem, and the actual expenses must be

Table 5.1 Blank Project Pricing Form

Project: ____________________ **Date:** ____________

	TASK	TIME Self	TIME Travel	TOTAL HOURS
1				0
2				0
3				0
4				0
5				0
6				0
7				0
8				0
9				0
10				0
11				0
12				0
13				0
14				0
15				0
16				0
	Total Hours	**0**	**0**	**0**
	Hourly Rate	$ 130.00	$ 65.00	
	Project Subtotal	**$ —**	**$ —**	**$ —**

	Other Direct Expenses	
1	Subcontractor	$ —
2	Airfare	$ —
	Per Diem	$ —
	Hotel	$ —
	Rental car	$ —
	Taxi	$ —
	Tolls	$ —
	Parking	$ —
	Meals	$ —
3	Misc	$ —
	Supplies	$ —
	Printing/Copying	$ —
	Telephone	$ —
	Subtotal Project Expenses	**$ —**

Project Pricing	
My Labor	$ —
Other Direct Expenses	$ —
Subtotal	**$ —**
Profit @ 10%	$ —
PROJECT TOTAL	**$ —**

Table 5.2 Project Pricing Form with Hours

Project: ABC Company **Date:** 1/26/06

	TASK	TIME Self	TIME Travel	TOTAL HOURS
1	Site visit	8	2	10
2	Design	3		3
3	Report	6		6
4	Design Specs	16		16
5	Drawings	3		3
6	Pre-bid conference, questions	5	2	7
7	Evaluate bids, approve shop drawings	3		3
8	Check job progress	12	6	18
9	Change orders	2		2
10	Test	4	2	6
11	As-builts	2		2
12	Contractor payments	2		2
13				0
14				0
15				0
16				0
	Total Hours	**66**	**12**	**78**
	Hourly Rate	$ 130.00	$ 65.00	
	Project Subtotal	**$ 8,580.00**	**$780.00**	**$9,360.00**

	Other Direct Expenses	
1	Subcontractor	$ —
2	Airfare	$ —
	Per Diem	$ —
	Hotel	$ —
	Rental car	$ —
	Taxi	$ —
	Tolls	$ —
	Parking	$ —
	Meals	$ —
3	Misc	$ —
	Supplies	$ —
	Printing/Copying	$ —
	Telephone	$ —
	Subtotal Project Expenses	**$ —**

Project Pricing	
My Labor	$ 9,360.00
Other Direct Expenses	$ —
Subtotal	**$ 9,360.00**
Profit @ 10%	$ —
PROJECT TOTAL	**$ 9,360.00**

itemized, but this is rare, especially when the majority of projects are billed on a firm fixed up-front price.

When the consultant bids a project, the airfare expense must be estimated in the best manner possible. Some consultants use Internet services such as Orbitz and Travelocity to mimic an actual trip and come up with cost estimates and schedules. Others deal directly with an airline's website to estimate the cost. In either case it is recommended that the estimated cost be based on a flight with little advance timing. It is never known for certain how much time a consultant will have between being awarded a project and when it must begin. If the airfare cost is based on a longer advance time, the consultant may take an unwanted loss.

For projects that do not include airfare but instead require consultants to use their own cars, some consultants charge mileage. The IRS has set standards as to mileage rates, or consultants can do whatever they want if the client has no policy and there is a fixed project price. In many cases this mileage is not figured or billed; instead, for local projects, those costs are lumped in with the travel costs in the marketing and advertising budget used to figure the hourly fee.

Miscellaneous

Miscellaneous expenses include printing and copying costs, supplies, and telephone costs that can be attributed strictly to the client and are outside the normal costs in these categories that are included in the office supplies or communications sections of the budget used to figure the hourly fee. Usually, a somewhat substantial cost is included here; minor expenses would be part of the budget number. Of course, any expenses unique to the particular project that have not been categorized can be included within Miscellaneous.

Table 5.3 shows a fully completed Project Pricing Form. The expenses have been included and totaled in the appropriate field. An additional line item is called Profit. Although company profit has already been calculated in the hourly fee, it may be prudent to add some extra money in case expenses rise between the time the project is bid and when the work takes place, especially with the sporadic cost of airfare and gasoline. The percentages can certainly vary as well as the totals the percentage is taken from. This example shows a 10% addition just to the expenses, not the consulting fees.

A similar form, or the same form with actual figures added, can be used to keep track of the actual time it took for each task and the actual expenses for the project. When a project of similar size and scope is bid on, the consultant can go to the completed project form and compare what the

Table 5.3 Completed Project Pricing Form

Project: ABC Company **Date:** 1/26/06

	TASK	TIME Self	TIME Travel	TOTAL HOURS
1	Site visit	8	2	10
2	Design	3		3
3	Report	6		6
4	Design Specs	16		16
5	Drawings	3		3
6	Pre-bid conference, questions	5	2	7
7	Evaluate bids, approve shop drawings	3		3
8	Check job progress	12	6	18
9	Change orders	2		2
10	Test	4	2	6
11	As-builts	2		2
12	Contractor payments	2		2
13				0
14				0
15				0
16				0
	Total Hours	**66**	**12**	**78**
	Hourly Rate	$ 130.00	$ 65.00	
	Project Subtotal	**$ 8,580.00**	**$780.00**	**$9,360.00**
	Other Direct Expenses			
1	Subcontractor	$ 1,070.00		
2	Airfare	$ —		
	Per Diem	$ —		
	Hotel	$ —		
	Rental car	$ —		
	Taxi	$ —		
	Tolls	$ —		
	Parking	$ 20.00		
	Meals	$ —		
3	Misc	$ —		
	Supplies	$ —		
	Printing/Copying	$ 150.00		
	Telephone	$ —		
	Subtotal Project Expenses	**$ 1,240.00**		
	Project Pricing			
	My Labor	$ 9,360.00		
	Other Direct Expenses	$ 1,240.00		
	Subtotal	**$10,600.00**		
	Profit @ 10%	$ 124.00		
	PROJECT TOTAL	**$10,724.00**		

estimates were versus the actual numbers. The bid for the new project should then be closer to reality, making for a better proposal.

FINAL PROJECT COST

Once the Project Pricing Form has been completely filled out or other method of determining project cost has been completed, the total project cost needs to be analyzed to make sure this is the price the consultant wants to charge the client. In many cases, the price is what it is because all the correct steps have been taken to correctly estimate the project cost. However, after the consultant looks at the total price, it may be appropriate to make some alterations for the following reasons:

- The total price comes to $10,040. Changing the price to $9,080 is probably a good idea. Even though it's only $60 difference, the psychological difference is far greater. Crazy but true.
- If the consultant is short on work and temporarily wants to bid more competitively to be busier, altering the price may be appropriate, especially for a brand new consultant. The emphasis here should be on *temporarily*; to continue this practice would mean never reaching the salary goals set at the beginning of this exercise. Bidding lower to enter a new market or geographical area might be similarly appropriate.
- Sometimes consultants, just like all businesspeople, get a bad feeling about a client or project—not enough to walk away from the project and not bid at all, but enough of a bad feeling to be cautious and concerned. Bidding the project a little higher in case bad or unforeseen things happen during the project may be appropriate. Every consultant will eventually have a project that goes bad to some degree; it would be nice if such circumstances were foreseen and planned for. In most cases, that will not be the case.

After this analysis is complete, the consultant can add the project price to the proposal and see what happens. Everything else being equal, if the project has been priced correctly, a certain percentage of project proposals will be accepted and the work can begin.

BILLING

Actually getting paid for services is a critical element of the consulting process. A method of getting paid needs to be established between the consultant and client that is acceptable to both parties. Sometimes this is a negotiating point, and other times it is a set process per the client. Whatever the case, the exact method and timing of payment needs to be agreed upon and understood before the project begins. Failure to do so might create headaches and hard feelings as the project progresses and the consultant is looking for some payment.

For projects that are billed on an hourly basis plus expenses, depending on the size of the project, some advance may be requested by the consultant before beginning work, at least to cover all or a portion of the expenses. Not all clients are willing or able by company policy to give an advance; this is part of the contract negotiations. If the project will last over at least a several-month period, billing is typically monthly based on the hours of work performed and the expenses incurred that month. If there was an advance, that amount needs to be deducted from the appropriate number of monthly bills. If the project is short in duration, even if spanning over the first of a month, one final bill may be appropriate. Net 30 Days is a standard billing practice although requesting quicker payment is certainly not taboo. Based on client policy, quicker payment may not happen, but there is no harm in asking.

For projects that are billed on a firm fixed up-front price, a consultant first must find out whether the client uses a purchase order (PO) system for payment. A purchase order is basically an accounting tool used by purchasing departments to verify that monies are set aside for payment against the project and that invoices should be paid against the purchase order after some sort of verification process or sign-off that the work has taken place. If the client uses a PO system, advances are out of the question. If the consultant wishes to bill monthly against the PO based on the amount of work done and expenses incurred during that month, those arrangements need to be made and verified with the purchasing department or client contact before the project or billing begins. Most POs are paid on a Net 30 Days basis. Otherwise, the full amount of the PO is paid at the end of the project. Private client POs are an excellent method of guaranteeing payment, assuming the work is performed to the client's satisfaction. Municipal POs are as good as gold and a definite guaranty of being paid and probably on time. If a client uses a PO system, it is important that the consultant has a copy of the PO form with a PO number on it before work begins. In essence, the PO works as a contract between the consultant and client.

For firm fixed up-front price projects in which the client does not use a PO system, it is recommended that an advance be requested. The advance ensures that the client is serious about the project and paying for it; plus, it covers the consultant's expenses so they don't have to be fronted until an invoice is paid. Although the requested advance percentage varies from

25–50%, 33.3% ($^1/_3$) is probably the most common. The balance can then be billed on a monthly basis, on a scope of service basis, or at the end of the project, depending on the arrangements made between consultant and client. If the client is unwilling to pay an advance under these circumstances, that is a red flag. This unwillingness may be a sign that being paid from this client is going to be a problem.

The Invoice

With today's technology, most invoices are prepared in some sort of software program, in many cases associated with an accounting program. They can also be done in stand-alone software. The invoice formats can be customized to the consultant's liking and particular billing needs. Invoices for projects billed hourly will be much more detailed than for projects on a fixed price because descriptions of the tasks performed with the associated costs need to be spelled out. Each task and expense needs to be itemized separately. For fixed-price project invoices, that kind of detail is not necessary; a summary of work performed with an associated cost and a list of expenses is sufficient. Figure 5.1 shows a sample invoice from an accounting piece of software with moderate detail.

Billing Schedule

Having to say so may seem silly, but invoices should be sent out in a timely manner. If monthly payments are being requested, then the invoice should be sent on time at the beginning or end of the month. If the project is being billed when completed, then the invoice should be sent within a reasonable amount of time after the completion of the project. It is wrong to expect payment to be on time if the consultant isn't conscientious enough to send the invoice on time. Being busy or not liking to do the paperwork is no excuse. Besides, cash flow will come to a screeching halt if billing is not done in a timely manner, and cash flow is the lifeblood of a business.

Late Payments

Invariably, some clients will be late paying their bills; it's the nature of business. If a client is on a PO system, a call to the purchasing department politely requesting to know when payment will be forthcoming almost always takes care of the problem. At worst, the consultant will find that the invoice hasn't been signed off by the appropriate person, so he can deal with that issue. Otherwise, the consultant will receive a firm date for

Strategic Design Services, LLC

P.O. Box 436
Portland, CT 06480

Invoice

Date	Invoice #
8/22/2006	ABC 1

Bill To
ABC Company 1234 Main St. Anywhere, USA 12345

P.O. No.	Terms
987654	

Description	Amount
Security & Access Control System at Company Headquarters Agreement number 00-0001 Project number PPP-001	
Procurement phase per proposal and service agreement	
Consulting Services	4,590.00
Airfare	525.00
Car rental and parking	150.00
Per Diem - 2 days	470.00
Total	$5,735.00

Phone #	Fax #	E-mail	Web Site
860-342-2544	860-342-0542	BDGStrategic@aol.com	Strategicdesignservices.com

Figure 5.1 Sample Invoice

payment most of the time. If the client does not work on a PO system, there are some options to help the process along. The best and most effective is probably a phone call or personal e-mail to the client contact asking if there is an issue and for an update on payment. Another option is to send statements after 30 days have passed as a reminder. Some consultants also add a small late fee percentage to late invoices. However, this policy needs to be spelled out in advance of the project work; otherwise, that charge will almost certainly not be paid.

If the work is not completed to the client's satisfaction and therefore an invoice is not being paid, the consultant must try to work out the problem with the client. There is no magic solution to this problem, but ignoring it certainly isn't the right course of action. If the quality of the consultant's work is up to par, any discrepancy should be able to be worked out. If the work is not up to par, the consultant needs to do whatever is required within reason to make it the quality it should be. Not doing so will make getting paid more difficult and will cause great harm to the consultant's reputation that will be difficult to overcome. If the work is completed to the client's satisfaction but the client simply won't pay the bill, then anything from phone calls to letters to legal action in small claims court are options. An attorney should definitely be consulted if the problem becomes this severe. These drastic measures are fortunately rare, and the hope is that they will never happen in a consultant's career.

6 Assessment

The first group of services provided by a security design consultant is assessment services, which are provided in some form for all projects. A physical security countermeasure system cannot be designed or specifications written without knowledge of the specific requirements of the system, and that information cannot be known without some form of assessment. Project management services are not provided unless there is a designed system implemented first. Assessment is also the only group of services that can stand on its own. An assessment can be performed and report written without the consultant's ever taking the next step of design. This may be intentional from the beginning of the project or happen because of the results of the assessment.

ASSESSMENT SERVICES

There are a multitude of different assessment services that a consultant can provide and that a client will request. No two projects are alike; they all have their variations based on the client requests, the particular circumstances of the project, and the individual methods of the consultant. Following are some general categories of services that can be provided:

1. If the type of physical security countermeasure required or requested for the project is known up front, the assessment can make the determinations about the design criteria of that countermeasure so the system can then be designed.
2. If the type of physical security countermeasure desired or needed is not known up front, the assessment can be the evaluation of the facility to determine what physical security countermeasures are recommended and then determine the design criteria of those systems.
3. There may be an assessment of existing physical security countermeasures at the facility both in terms of the functionality of the systems and whether the existing systems meet the current or future needs of the client.
4. There may be an assessment in regards to code compliance of existing physical security countermeasures such as fire codes and Americans with Disabilities Act (ADA) codes. Code knowledge will be discussed later in this chapter.
5. A report in most cases is the result of the assessment detailing the findings of the assessment and the recommendations of the consultant. Details of the report will be discussed in Chapter 10.
6. A security design consultant can also make a physical security assessment of a facility as part of a larger security assessment with a security management firm.

All assessment services for which the facility actually exists require a site visit. The consultant has to see a facility to make any recommendations or gather the necessary information to design any physical security countermeasure system. Not to do so and instead rely on other people's descriptions or evaluations is unprofessional and will not produce the quality end result expected from a security design consultant. In some cases in which similar systems are designed in multiple facilities of the same type for the same client, not every facility is visited, but in those cases at least one facility has to be visited for the consultant to perform the initial assessment for the designed system. From there, drawings can be relied on to access the facility. Also, if the facility doesn't yet exist and drawings are the only tool to work with, visiting the facility is obviously impossible. There are even less common cases in which the client has all the products and device locations predetermined and just wants the design specifications written and possibly project management services provided. Even so, the responsibility of the security design consultant is to have some exchange with the client to make sure those decisions are the correct ones and in the best interest of the client.

ASSESSMENT TOOLS

A consultant may require some specific tools while performing an assessment for a physical security countermeasure system. Which tool or tools depends on the individual facility and countermeasure in question. The consultant should obtain the necessary tools needed based on the specific services offered. It is a good idea to put together a toolkit that can be brought along to any assessment project. Having a toolkit is easier than trying to remember everything that is needed for the project being worked on. The toolkit also should be travel friendly.

Necessary tools can include the following:

1. *Flashlight:* In many cases a security design consultant will have to look inside closets, cubby holes, cabinets, consoles or above ceilings, etc., to perform an assessment. Having a flashlight on hand is much more professional than asking the client to borrow one. One of the small Mag-Lite flashlights that manufacturers like to give away as a promotional item is usually sufficient and fits nicely into a pocket when walking around. A flashlight is almost always needed during an assessment.
2. *Measuring Wheel/Tape Measure:* A security design consultant is constantly taking measurements during an assessment, whether in a room or parking lot. For measuring longer lengths, a measuring wheel is a great tool. Because these wheels are used mostly by surveyors, a surveying shop is the most likely place to buy one; these wheels store compactly but extend out so the consultant can just walk along and measure underneath them. For smaller lengths a simple tape measure is used and should be with the consultant for probably every assessment.
3. *Digital Camera:* The advent of digital cameras has added an extra dimension to the consultant's abilities. First, pictures of actual areas of the facility or particular details can be input into the computer and included in the consultant's report. A picture is worth a thousand words. Also, pictures can be taken during an assessment to help the consultant remember what something looks like and can capture details that the consultant might not otherwise take into account when designing the system.

4. *Focal Length Finder/Lens Chart:* When the consultant is making a determination as to what size lens or range of lens is appropriate for a specific camera, a focal length finder is very helpful. This device mimics the view of the camera and is adjusted until the desired field of view is seen. Then the corresponding focal length is read from the finder. In lieu of this device, lens charts can also be used, although measurements must be taken before using the charts. Both of these items can be bought from a CCTV manufacturer.
5. *Hand-Held Display Monitor:* To see the field of view for existing cameras if a monitor is not present or is broken or if wiring is damaged, the consultant can use a hand-held display monitor at the camera location. This tool is also very helpful during the project management stage. This type of monitor can be purchased at an alarm equipment distributor.
6. *Light Meter:* If lighting recommendations and findings are part of the consultant's offered services, a light meter is a must. It measures the light in an area, giving a specific numeric reference used against industry lighting standards. Light meters are easily found on the Internet for purchase.
7. *Small Hand Tools:* Small tools such as screwdrivers, Allen wrenches, and wire tags are good to have available in case they are needed. If a security cabinet needs to be opened or pull station needs to be examined, it's a time-saver if the tool for the job is readily available.
8. *Sample/Dummy Products:* Manufacturers make and give to consultants sample and/or dummy products such as access control readers, key tags and cards, keypads, cameras, etc. Even though the security design consultant is *not* trying to sell any particular product, these samples can be very helpful in showing the client the options available to get feedback. It's no use recommending access cards if the client wants key tags.
9. *Testers/Keys:* There are different test devices for products and specific keys for different manufacturer cabinets. If the consultant comes from the integration world and therefore has some of these items, they could come in handy. If not, it's not a big deal in the vast majority of assessment projects.
10. *Drawings/Graph Paper:* While drawings are not technically tools, the need for drawings of the site facility cannot be overstated. They help navigate the facility, and notes can be written right on the drawings for future reference, which is more efficient than describing the location on a notepad. Measurements can also be

noted right on the drawing. In certain instances graph paper can be used instead to draw a particular area although that approach would be very time consuming for a whole facility.

11. *Recorder:* Small digital recorders are a great help during an assessment. Thoughts that the consultant has any time during the process can be spoken into the recorder for later listening and action. Many people have great thoughts at times when they can't write them down and then forget them later. At least, this has been a problem for this author. The recorder helps solve that problem. In addition, some recorders have the capacity to record longer interviews, so note taking isn't as important and the consultant can actually listen to the interviewee better.

INTERVIEWS

Almost every assessment begins with interviews in one form or another. Information must be gathered about the project and the facility from the people who are or will be interacting with the physical security countermeasures in the end. These are the people who are dealing with the problem and will benefit from the solution. Following are some of the overall categories of information that need to be gathered from the interviews:

- Project budget
- Project schedule
- How the project fits into company or construction operations
- Desired outcome of any installed system
- Company culture
- Future plans for security
- How things actually operate right now
- Impression of the company's security

Not every subject will be discussed with every person interviewed. For instance, the project budget would not be discussed with a random employee. The interviews need to be tailored to the person being interviewed, the type of client, and the type of physical security countermeasure system or project type.

Who to Interview

As a general rule, the consultant needs to interview the *project stakeholders*—in other words, those people, or a representative from a group of people, who will be affected by the installation or noninstallation of the physical security countermeasure system. There is no point in interviewing someone who is not affected at all by what is being done, except if doing so is politically necessary within the company. Likewise, if someone who is affected by what is being done is not interviewed, the assessment will not be as effective and of the best quality; therefore, the physical security countermeasure installed may not be the best fit for the client's needs. Some people who commonly are interviewed include

- *Main Contact of Consultant for the Project:* This person has had the responsibility of working on the project to this point and will certainly have a lot of information to offer.
- *Chief Executive Officer (CEO), Chief Operations Officer (COO), President, Chief:* This person may or may not have a lot of good information but will know company culture and in many cases needs to feel part of the process.
- *Director of Security:* For obvious reasons, a security design consultant should always interview the security director.
- *Construction Manager:* If the project is part of a larger construction project, the consultant needs to gather all kinds of information to make sure the system fits in with the rest of the project.
- *Custodian/Janitor:* Interviewing the custodian may sound like an odd choice, but who knows the facility better? Who knows the comings and goings of employees and how the facility opens and closes better than the custodian? The custodian also tends to tell it the way it is.
- *IT Department:* This or any department that will have some interaction with the installed physical security countermeasure should be interviewed.
- *Other Contractors:* If the project is part of a larger overall project, any other contractor that may be interacted with during the installation might be interviewed.
- *Sampling of Random Employees:* Because most of the others being interviewed are management, it is a good idea to also interview a small random sampling of employees. They tend to know more of what actually happens than sometimes management does.

The consultant also should ask the company contact who he thinks should be interviewed. This person will certainly have a better idea of who the project stakeholders are than the consultant.

Telephone Versus Personal Interviews

Should the interviews be done in person, or can they be done over the telephone? In-person interviews are always best for the same reasons that in-person initial client meetings are best. The rapport and give and take of face-to-face contact will almost always make a better interview. There are, however, other considerations. Are the people who need to be interviewed available during your site visit? Can those interviews be scheduled on site with the help of client personnel? How many extra days are necessary for the site visit to conduct the interviews, and is the extra travel expense cost effective for the client? In many cases it is necessary to have the interviews over the telephone before the site visit. The consultant should make sure they are done before the site visit so any information gathered can be confirmed or addressed during the site visit.

Interview Questions

Before the interviews begin, the consultant should write a questionnaire with all the questions that need to be asked to serve as a guideline and make sure nothing is forgotten. Because different questions need to be asked to different people, the consultant can write separate questionnaires or just skip over the inappropriate questions during the questioning. A separate questionnaire sheet should be used for each interview so the answers can be recorded on the specific questionnaire and attributed to the correct person. In some cases the answers to questions are included in an assessment report.

It is impossible to provide exact questions that would be used in a questionnaire because they are dependent on the individual project and scope of work. The following questions are in some cases generic and nonspecific to any physical security countermeasure but give a range of the types of questions that need to be asked. The actual questions must be very specific to get an accurate response and keep the interview on track. One of the following questions may, in fact, encompass many questions on the actual questionnaire. These questions include, in no particular order, the following:

1. How much money is budgeted for the installation of the physical security countermeasure? Are any ongoing costs budgeted?
2. What is the schedule for the installation of the system?
3. Is this project part of a larger overall construction project? If so, how will this installation interact with the rest of the construction?

4. How will you be interacting with the installed physical security countermeasure?
5. What is the end result you would like to see from the installation of the physical security countermeasure?
6. What end result would you *not* like to see from the installation of the physical security countermeasure?
7. Was there a particular event or occurrence that triggered this project?
8. What are the security problems of the facility?
9. Is there a particular physical security countermeasure that you believe is needed? Is there one that is not needed and why?
10. What are your concerns about the security of the facility? (This question is different from #8 and asked of different people.)
11. What are the expectations for the level of security at this facility?
12. How can the security of the facility be improved?
13. What are the threats to the facility?
14. How is the facility vulnerable from those threats?
15. How will employees react to the installation of a particular physical security countermeasure?
16. What departments would be involved with the installation? What would their roles be?
17. Are there any future plans other than this project for the security of the facility?
18. What are the existing physical security countermeasures in place?
19. How do the existing physical security countermeasures work and operate? Are they effective?
20. What is the daily routine of the facility?
21. How is the facility opened and closed for business?
22. What personnel type and number will be dedicated to the use or administration of any installed physical security countermeasure system?
23. Do you have any other observations or suggestions?

Getting any information out of some people will be difficult no matter how specific the question, and others will go on at length if they are allowed. The different responses will not necessarily be a reflection of what information a person has to offer. Communication skills come in handy during interviews. There are interviewing methods that teach how to "drill down" when asking questions in order to obtain the necessary information. Education in interviewing techniques is not a bad idea if the consultant has some difficulty in this area. The length of each interview certainly varies with the person being interviewed and the scope of the project, but each interview generally should not go over one hour. Some will naturally be shorter, and some will have to be longer. The information gathered from the interviews is then used to help provide the assessment results.

OBSERVATIONS

The next step in the assessment process is site observations—not just looking, but observing. A tremendous amount of information is available by just observing what is going on within a facility. Walking the facility once to verify conditions or doing a survey of assets and vulnerabilities doesn't go far enough. This is a crucial step and one that unfortunately not every consultant utilizes. A good work product may still result if this step is not undertaken, but the best possible work product certainly won't be. It also happens to be the favorite element of the assessment for this author, who found the following information for just one client from observing:

1. While observing the operations of a police department client by just standing in the corridor for a half hour or so, this author noticed that officers were leading people under arrest in handcuffs through the same corridor as civilians were allowed to be in to pick up evidence. This information was never mentioned in interviews and would have been missed by just walking the facility. The problem of this situation occurring would not have been addressed if not for the observation.
2. All the exterior doors of this PD facility were going to have access control readers installed on them. However, during observations, this author noticed that one exterior door to the basement boiler room was open. By going back to that door several times during the day, this author observed that door was open quite a bit to allow contractors access to the boiler room. From that piece of information and the subsequent questions that were asked to the client, it was determined the reader should go on the interior door from the boiler room to the rest of the building. If that door had not been open at the time of a simple walkaround, the system would have been designed improperly, or at least not in the best manner possible.
3. While observing, this author noticed that one particular area surrounding an exterior door was being used by the officers as a gathering point to exchange information about the day, etc., and the door was propped open for periods of time while these

> conversations took place. It needed to be verified that this door should still have a prop-open alarm and that behavior, which was really ingrained into the officers' routine, would no longer be able to take place.

The findings from this observation were not a fluke; important information relevant to the particulars of the physical security countermeasure to be assessed or designed is always found through observation. The consultant should make sure enough time has been allocated in the project budget and schedule for this step. The first observations can certainly be in conjunction with the site survey described in the following section, but time for additional observations should be allocated. Visits to the facility for observations should be made at different times of day and night. Different observations will be made during the night than during the day. If a facility has multiple shifts, observations should be made during each shift. The total number of observation visits and total time allotted for observations will vary per project but should be enough to thoroughly observe all pertinent facility operations. As the consultant becomes more experienced at observation, he will develop a feel for what is enough.

SITE SURVEY

The traditional form of gathering information during the site visit is the security survey. Whole books are dedicated to all the categories of information and specific information to be gathered during the security survey. The security survey is a well-used and valuable tool for the security management consultant but isn't as well suited for the security design consultant. Only the physical security section of the traditional security survey would be used, and the questions are usually too generic to help with an actual system design. That is not to say that a security survey can't be written by an individual consultant based on specific needs. However, the traditional security survey form is not used by most security design consultants.

Instead, what is done might more appropriately be called a site survey. Evaluations are made of the facility during a complete walk-through and previously described observations, paying attention to three key areas: assets, threats, and vulnerabilities.

Asset Identification

The assets that need to be protected with the physical security countermeasures must be identified. In some cases the client will already have told the consultant what are the particular assets of concern that need to be protected. Even so, the consultant should look for other assets of note and at least bring them to the attention of the client. In other cases the consultant is starting from scratch because the client simply wants the facility protected. Physical assets like inventory and office equipment are the most obvious that need protecting. They should be identified and their locations noted. Assets also include people, both employees and visitors. How people enter and exit the building and their movements within the building need to be analyzed. How they get from their vehicles to the building is also of key concern. Information can also be an asset, such as the proprietary formula for the client's product.

Personnel and/or policies and procedures countermeasures may need to be integrated with physical security countermeasures to protect the asset. The security design consultant deals only with physical security countermeasures but must have enough expertise to realize when the services of a security management consultant are also needed on a project. The fact that this happens quite a bit is why the two disciplines of security consultants partner frequently on projects.

Threats

While the evaluation of external threats to a facility based on previous occurrences and the surrounding area is the domain of a security management consultant, obvious threats can be identified by the security design consultant during the site survey. Internal theft and vandalism are good examples. Also, threats identified through the interview process can be verified. Common sense also plays a part; if the facility is a medical testing lab that uses animals, it's a safe bet that People for the Ethical Treatment of Animals (PETA) should be on the threat list.

Vulnerabilities

The real expertise of the consultant comes into play and the majority of the site survey time is spent trying to identify vulnerabilities. What are the vulnerabilities of the facility in protecting the assets from the threats? Every area of the facility needs to be examined from a security point of view to evaluate the vulnerabilities of the facility. Through experience, the consultant develops the expertise to look at a facility and identify the security weaknesses. These identified vulnerabilities will be used to determine the physical security countermeasure criteria used to design an appropriate system to be implemented.

DESIGN CRITERIA

All the work completed so far in the assessment has been done to answer one question: What physical security countermeasures and/or design criteria for those countermeasures should be implemented to erase the vulnerabilities in order to protect the assets from threats? While design criteria are actually part of the assessment process, this topic will be explored in detail in Chapter 7 on system design.

CODE COMPLIANCE

Code compliance is an important issue with the assessment of physical security countermeasures as well as the design of those countermeasures. The government at all levels—federal, state, and local—has adopted codes, standards, guidelines, and ordinances relating to the design and installation of physical security countermeasures. The local Authority Having Jurisdiction (AHJ), which could be a building official, fire marshal, or other inspector, verifies that the design and installation of the countermeasures meet or exceed all the applicable regulations. Regardless of the physical security countermeasures chosen for the facility, it is essential that they meet or exceed the regulations that apply to the particular application. While this is important for all measures, it is particularly important for access control and fire systems.

A *code* is a government-mandated method by which a system should be designed and installed. *Standards* are also set policies of design and installation. *Guidelines* are less mandated policies than codes and standards but are also important. *Ordinances* are mandates at the local government level. A consultant should never recommend a countermeasure or the design of a countermeasure that goes against a regulation. Doing so shows a complete lack of expertise and leaves the consultant open to liability.

A consultant should always point out code deficiencies that are seen during an assessment. That information should not always be put in writing, because doing so may put the client in an unwarranted bad position. Putting in writing that access readers are not at ADA-required height is one thing, but putting in writing that the client's whole fire system is out of code is quite another, unless, of course, that is why the consultant was hired.

The security design consultant should have a good working knowledge of the applicable codes, standards, and guidelines based on his specialties. They don't need to be memorized necessarily, but well understood certainly. Following is a list of some of the applicable codes, standards, and guidelines and what they reference:

1. *NFPA 72, National Fire Alarm Code:* This code not only has fire system installation regulations but also includes integration of access control systems to fire systems.
2. *NFPA 101, Life Safety Code:* This code has a whole section on door requirements for access control, means of egress, etc.
3. *NFPA 70, National Electrical Code:* This code deals with correct wire sizes and types, power requirements, etc.
4. *IESNA G-1-03, Guideline for Security Lighting for People, Property, and Public Spaces:* This guideline has lighting standards to compare with light meter readings.
5. *ADA, Americans with Disabilities Act:* This act deals in particular with height of access control and fire system devices, fire notification device qualities, etc.
6. *UL Standards:* Underwriters Laboratories has standards and guidelines for numerous security devices and systems, including burglary, CCTV, access control, and gate operators.
7. *BOCA Building Codes:* These codes deal with mostly construction issues but include fire caulk for wire holes, core drilling, etc., that are necessary to be included in design specifications.
8. *Local Codes and Ordinances:* The consultant always needs to check with the local authorities to make sure there is no local code issue that needs compliance.

Many others may apply to a particular system or facility and need to be adhered to for a particular project.

7

System Design

The second group of services provided by a security design consultant is system design services. Designing a physical security countermeasure system is a three-step process. The first step is to collate and analyze all the design criteria gathered during the assessment phase of the project to determine what physical security countermeasure system or systems should be used and what all the elements of those systems are. In other words, what are the needs of the facility? All this information is usually included in the assessment report. The details of this step will be covered in this chapter. The second step is to match those facility needs to exact products. This information may or may not be included in the assessment report, depending on the scope of the consultant's work. The details of this step will be discussed in Chapter 8. The third step is to write design specifications and produce drawings for the designed system. The details of this step will be discussed in Chapter 10 on reports.

Designing a physical security countermeasure system begins with gathering all the design criteria during the assessment phase of the project to determine what physical security countermeasures are appropriate for the project and what elements or details of that system are needed for design. In other words, what are the needs of the facility? Before the information can be collated and analyzed, it must be gathered during the assessment site visit. If it is not known going into the assessment what physical security countermeasures system will be utilized, then this step is broken into two parts. The first part is making the determination as to what physical security countermeasure system or systems will be recommended, and the second part is finding out all the details required to design those systems. If it is known going into the assessment what physical security countermeasure system is requested and will be designed, only the second part will be required. However, even in that case, the consultant should verify—based on site conditions, expertise, and industry best practices—that the requested system is correct for the needs of the facility.

BASIC DESIGN GUIDELINES

Some basic design guidelines should be followed during the design process. The purpose of these guidelines is to ensure that the resulting system design is of the best quality possible. When a security design consultant says that best practices were used in the design of the system, these guidelines can be referenced to support that claim. They include the guidelines described in the following sections.

The Design Must Fit the Need

Ensuring the design fits the need is by far the most important concept in designing a system. It is a fundamental concept and may seem obvious, but it is surprising how many physical security countermeasure systems are installed in the field that do not meet the needs of the client. For instance, the client is concerned about who is entering the facility, yet the entrance doors have no access control system of any kind and the cameras are turned inward at the entrance points. Designing a system is not a cookie-cutter process. No two applications or facilities are the same, which means no two sets of needs are the same, and therefore, no two designed systems should be the same. Different systems may be similar, of course, but never exactly the same. Designing the same system over and over without much regard for the needs of the facility is lazy, unprofessional, and unworthy behavior for a consultant. The assessment phase of the project ended with the needs of the facility; now the design must meet those needs.

The Design Must Be Appropriate for the Facility

Many system designs may fit the needs of a facility, but the best system design must also be appropriate for the type and size of the facility in question. A 100-camera closed circuit television (CCTV) system for a 10,000 square foot facility might meet the client's needs but probably isn't appropriate. Having one of those cameras in a bathroom certainly is inappropriate. Likewise, having those cameras be pan-tilt-zoom (PTZ) cameras when no one is available to operate the controls is inappropriate, even though the client's needs are technically met. A security design consultant should not recommend more or less in terms of both quantity and functionality of system components than what is needed or appropriate for the facility. Again, the consultant's expertise and industry best practices come into play, as well as some good old common sense.

The Design Must Meet All Applicable Codes

Codes, standards, and guidelines were briefly discussed in the preceding chapter. The reason a security design consultant should have a working knowledge of these regulations is that the design must adhere to those regulations. Having an access control system card reader at a height of 50″ or only one fire system horn/strobe in a 100′ × 100′ room is just not acceptable. The client may or may not ever know if one of the first two guidelines was followed, but the client will surely know when the system does not pass municipal inspection or the contractor points out an error. If the consultant doesn't know for sure what the correct regulation is, he should pick up the phone and ask.

Applying these basic guidelines to a design distinguishes a security design consultant from an installation company product salesman when designing a system. The professional consultant has no financial or other incentive not to follow these guidelines. In fact, there is every incentive for a consultant's reputation and the quality of the work product to follow these guidelines. Therefore, it is in the client's best interest to hire the consultant to design the system.

INTEGRATION OF MULTIPLE PHYSICAL SECURITY COUNTERMEASURES

The basic function of a physical security countermeasure system is to detect, deter, and facilitate a response to a threat. Except for the smallest of systems, this function may require multiple physical security countermeasures. These countermeasures must be integrated together to form the entire physical security system.

Some manufacturers have already integrated multiple countermeasures into one functional system. For instance, some products currently on the market combine CCTV, access control, burglary, and even fire systems into one software package with compatible head-end equipment. The combinations of access control and CCTV or access control and burglary are the most common. Many of these systems are excellent, but the consultant should be careful, as there are pros and cons of designing this type of system that will be discussed in the next chapter.

The more common method of integrating physical security countermeasures is simply to provide multiple countermeasures and design them in such a way that they work in concert with each other. For instance, the detection can include a fence detector, CCTV cameras, outdoor motion sensors, a burglar alarm, and lighting. The deterrence can include fence razor wire, a gate operator, door locks, and access control devices. Facilitating a response can include forms of communicating alarms and an internal monitoring facility.

The consultant must first understand that the needs of the facility may have to be met with multiple physical security countermeasures. Different countermeasures have different functions and meet different needs. Then the consultant must know how to design the multiple countermeasures to form one functional system. In some cases the individual countermeasures will work totally independently, and in some cases they will need various forms of integration.

INTEGRATION OF PHYSICAL SECURITY COUNTERMEASURES WITH PERSONNEL AND POLICIES AND PROCEDURES COUNTERMEASURES

Physical security countermeasures, whether individual or multiple, are never used alone; they are always in conjunction with personnel or policies and procedures countermeasures or both. While this may be self-evident for large systems, even the most basic stand-alone burglar alarm has a procedure and/or policy for turning the alarm on and off, and personnel are involved if the system is monitored, whether internally or externally.

All electronic physical security countermeasures and most mechanical ones are not just installed and expected to work on their own. They require some kind of human interaction, in various degrees, for them to perform their functions correctly the way they were designed. Therefore, the personnel must not only exist to interact, but also must have an understanding of what to do. Here are just a few examples of that interaction:

1. An access control system cardholder attempts on multiple occasions to enter an area of the building for which she is not authorized and is denied access. The system software sets off an alert on the appropriate PC. The person in charge of monitoring and administrating the system must acknowledge the alert and determine how the situation should be handled. This scenario requires the personnel to monitor the system and the policy and procedure to determine what to do for every scenario.
2. A retail store has CCTV cameras throughout the store with PTZ capability to identify shoplifting and employee theft. Security personnel are watching the cameras and looking for criminal actions. When a suspicious activity is spotted, the personnel must determine the correct course of action and notify other appropriate security personnel. This scenario also requires the personnel, in this case specifically trained, to monitor the system as well as respond, and the policy and procedure to determine what to do.
3. There is a manned guard shack on a fence line with an automatic gate operator. Employees of the facility use an ID badge and card reader to open the operator and gain access, but visitors and deliveries must be checked in at the gate, and the operator must be opened by the guard for the person or vehicle to gain access. Not only must the guard exist, but there must be a policy and procedure in place for the guard to determine who is and is not allowed access.
4. An electronic turnstile at the employee entrance of a facility reads employees' ID cards and allows access. Inevitably, there will be employees who do not have or have lost their ID badges or, for whatever reason, are having a hard time getting through the turnstile. There must be some personnel in place and some procedure that is followed for these occurrences.
5. A guard force patrols the grounds and interior of a facility, but that requires adequate lighting for the guards to effectively see what is happening and an appropriate locking system for the guards to move around the facility and possibly a guard tour system to verify rounds. Additionally, the guards must have a set of policies to deal with situations that arise regarding the countermeasures.

The consultant must design the system keeping in mind that there will be interaction between the system and personnel or policies and procedures. Determining how, to what degree, and in what manner there is interaction between physical, personnel, and policies and procedures security countermeasures requires a degree of expertise and experience. It is also dependent on the needs of the client and the ability to provide personnel or develop policies. Regarding the design of a system appropriate for a facility, the client must be told what interaction is required for a particular system before it is designed to make sure the appropriate personnel resources are in place.

DESIGN CRITERIA

The design criteria need to be gathered and collated during the assessment site visit. The first part of gathering design criteria is determining what physical security countermeasures or combination of them will be used to meet the needs of the facility and how they need to interact with personnel and policies and procedures. To accomplish this, the consultant should ask and answer a series of questions, which include but are not limited to the following:

1. What functions do the physical security countermeasures need to perform to counteract the vulnerabilities? For instance:
   a. Is it necessary to have tight control over access to the facility or parts of the facility (this would be different for a retail store versus a corporate headquarters)?
   b. Is it necessary to have visual identification of people in the building or entering the building or the actions they take (again, this could be quite different for a retail store versus a corporate office)?
   c. Is it necessary to have tight control over access to the entire grounds surrounding the facility?
   d. Is it necessary to provide additional safety for employees or visitors, either within the facility or outside the facility?
   e. Does the facility require additional security deterrents after hours?
   f. Do specific areas of the facility require additional security measures? Do these requirements change based on the time of day?
   g. Does the location of particular inventory or equipment need to be tracked?
   h. Does the specific threat require special considerations (i.e., ballistic, chemical countermeasures)?
2. What number and type of personnel can or will be committed to the overall physical security countermeasure system? For

instance, if the facility can't or won't have personnel to continually monitor CCTV cameras, PTZ cameras would not be used. Will systems be monitored internally or externally?

3. What policies and procedures are acceptable based on the culture of the facility and its management? The consultant might not at this point, for instance, ask employees at a normal corporate office to have their persons and belongings searched every time they enter the building, whereas employees at a sensitive government office would be searched.
4. What makes common sense for the facility? Although it would be effective to have an armed guard posted all night at a secondary school, does that really make sense? By the same token, does it make sense to just have door locks and good lighting at a manufacturing facility?
5. What is the budget for the system? Although to some degree the needs define the budget, in the real world there are constraints, and there must be a convergence of meeting the needs of the facility and the money able to be spent. If the client has given no guidance on budget, again common sense must prevail.

Physical security countermeasures should be designed in layers with the underlying theme being a combination of measures to detect, deter, and initiate a response. Here are a couple of examples of the thought process:

1. If a medical research facility tests on animals, the threat assessment would have determined that radical groups like People for the Ethical Treatment of Animals (PETA) were a concern. In that case, among other things, it would be important to control access to the grounds of the facility as well as the facility itself, have special security measures in the specific research areas, provide specific policies for the safety of the employees, and have some sort of guard force in close proximity should an event be detected.
2. If a corporate office houses employees and a data center with sensitive information, among other things, it would be important to control access to the building itself, provide special security measures for the data center, provide some surveillance for the safety of the employees, and have a burglar alarm with internal or external monitoring after hours.

A security design consultant must be able to assess the answers to the questions posed here as well as many other questions specific to the project and client and, from that, determine the physical security countermeasure needs of the facility. The results of these findings, along with an explanation of the rational thought process that achieved the results, can be included in the assessment report.

DESIGN DETAILS

The second part of gathering design criteria is determining and recording all the necessary details required to design the systems. These are actually the nuts-and-bolts details required to pick the right products for the system. They include the following:

1. *Location of Devices:* Where will all the devices be located? This does not mean general areas but very specific locations, such as an access control reader 10″ to the left of the nonhinged side of door #001, 42″ top height above the floor. These determinations need to be made for every device because conditions change from place to place within a facility.
2. *Head-End Locations:* Where will controllers, power supplies, recorders, monitors, panels, etc., be located? Again, this information is very specific—not just in a particular room, but on the south wall of the first floor electrical room 10′ from the left corner or on a shelf 5′ above the floor next to the desk in office #001. If the security design consultant doesn't make the decision, someone else will, and it may not be the best decision.
3. *Device Particulars:* Color of the device, field of view, desired size, finish, etc., all need to be determined. In many cases it is not possible to make these findings after the site visit because all the conditions of the surrounding area can't be remembered accurately. It is necessary to be there looking at the place where the device will be installed to detail the particulars.
4. *Head-End Particulars:* Required size of cabinets, monitors, recorders, etc., also need to be determined—not only where will they be, but how will they be mounted, wired, located in relation to each other, etc. As with the devices, these specifics will not be remembered later and need to be determined at the site visit.
5. *Environmental Considerations:* Are there either internal or external environmental considerations that need to be taken into account in the design? These considerations may include external weather, contact with chemicals or gases, and internal facility temperature.

6. *System Functionality:* How do all the system components need to work together? What is the best method of wiring the system? What are all the functions that the system must perform to meet the needs of the client. To say a CCTV system is required isn't enough. The camera capacity, whether it is network based or not, its storage capacity, and its recording rate are just a small number of the additional details required.
7. *Dimensions:* Dimensions for everything are always needed at some point in the design process—for rooms, parking lots, doors, driveways, and spaces of all kinds. If something is in question, it should be measured and the measurement recorded. Here, a set of drawings for the facility is a big help.
8. *Facility Construction:* Are the walls cement block or drywall? Is there a drop ceiling or finished ceiling? Are there wire chases, access panels, or crawl spaces? Basically, how does a wire get from point A to point B? Maybe it doesn't, which also becomes a critical design element. This analysis has to be done in exhaustive detail. It's one thing to get the color of a reader wrong, but it's quite another to have the installation company not be able to get a wire to a reader.
9. *Other Existing Conditions:* The door hardware on the appropriate doors, the thickness of the doors, whether there is a network within the facility, whether it is wireless capable, and whether a generator back-up for power is available are just a few of the existing conditions that may need to be known. The particular type of physical security countermeasure and facility will dictate exactly what must be determined; there is no magic list.

Now the security design consultant has all the information needed to determine what products to specify for the physical security countermeasure system and, in turn, to write the design specifications and drawings.

REQUEST FOR PROPOSAL VERSUS INVITATION FOR BID

Design specifications can be provided in two different manners, reflecting two different types of bidding processes. Which manner of specification that is provided based on the type of bid will determine whether exact product choices need to be made at this point in the design process.

The first type of bidding process is a Request For Proposal (RFP). With an RFP, the exact products are not included within the specifications, but instead the criteria or specifications for the products within the systems and the systems themselves are detailed. The bidder then chooses a set of products that meet the specified criteria and submits a proposal outlining the products chosen and how they meet the requirements as specified, as well as costs for the system and other requested information. The client, with the help of the consultant, then evaluates all the submitted proposals and determines which best meets the specified criteria at the best cost and what is the best fit for the client's needs.

This type of specification puts the onus on the installation contractor to make sure every product chosen works properly together, and together they make a system that performs to the specified requirements. If they don't, it is the contractor's problem in theory. It also potentially relieves some liability from the consultant for errors. However, the consultant must concur that the proposed system meets the specifications and requirements before the system is accepted. The client or architect, if one is involved, may also prefer this form of specification. When the low bid is required to get the work, like most municipal and government bids, this method can also help to weed out the unqualified contractors.

The second type of bidding process is an Invitation For Bid (IFB). With an IFB, the exact products, including make and model numbers, are included within the specifications. The bidder then submits a bid, giving the cost to install the system as designed as well as other requested information. As long as the bidder is qualified to perform the project work, usually the low bid wins. One of the keys to this bidding style is to make sure the contractor qualifications are written very well to ensure the project is not stuck with a bad contractor. A low bid system has a bad reputation, particularly with municipal work, for a good reason: the quality of the work can be garbage. However, if the specifications are written clearly and tight and the contractor qualifications are appropriate, the low bid being accepted is not as common a problem. There are bad contractors that will be selected for projects whether or not they were the low bidders; that's just a reality of the industry. The consultant's job is to take the steps to make that happen as little as possible.

One concern clients may have, particularly municipal and government clients, is they are required by law in some cases to not be locked into one manufacturer during the bid process. Therefore, they may have reservations about using an IFB. An IFB can still have "or approved equal" language after the specified products and allow for a process in which the bidder can submit alternative products for approval. As long as the proposed substituted products meet the exact specifications of the specified products, they can be accepted. The particular language within the specifications regarding substitutions determines how strict or loose the substitution process will be. Once a client understands how it would work, using an IFB should not be a problem.

Another plus to having the bid process be an IFB, if the specifications are written precisely and correctly, is that the contractors like and appreciate it. Now, it is not the job per se of the consultant to make the contractors happy. In fact, many times it will be just the opposite. However, if the contractor has *all* the facts of the system up front, including the details of how each product works and needs to interact with each other to make a properly working system, the procurement and construction phases of the project will go much smoother and will therefore make the client happy, which is the consultant's job. The more details are determined and decided by the consultant rather than the contractor, generally the better off the project will be. That is not to say that some good contractors will have important contributions to the process; indeed, they will. It's just that their specialty is installation, and not necessarily quality design as defined by a professional security design consultant.

Regardless of whether the form of bidding is an RFP or IFB, the consultant has to go through all the design steps as described in this chapter. Even if the exact products are not named within the design specifications, the end result of how the system is designed to work needs to be determined and included. That requires all the design steps to take place. In fact, in some cases the only discernable difference between an RFP and an IFB should be the writing of the product make and model numbers.

There is no right or wrong manner to provide specifications. There are excellent security design consultants who provide specifications in both forms. The choice of which to use is, to a large degree, personal preference, although the preference of the client may also come into play. This author almost exclusively provides specifications with product choices for an IFB. It is a selling point to the client that the consultant has the expertise to design the system with all the correct products that will work together in a manner that meets the needs of the client. Anecdotally, most clients prefer this method of bidding and specifications if they are given the choice. The key in either case is to write quality specifications.

The final step in the design process is to determine the products that will be specified to meet the design criteria and details as determined during the site visit and assessment.

8

Determining What Products to Specify

The final step in the design process is determining what products or equipment will make up the designed system. All the hard work put in by the consultant up to this point will be for naught if the wrong products are chosen. It won't matter that all the design criteria established during the assessment process were correct; the system will not function as it was designed or as it should. This step needs to take place whether the bid is in the form of an Invitation For Bid (IFB) or Request For Proposal (RFP). If it's an IFB, the exact products and equipment used for the system will be chosen. If it's an RFP, a potential set of products must be chosen to put the correct product criteria for the system, as described through Architects and Engineers (A&E) Specifications, within the design specifications. As stated previously, sometimes the only difference between an IFB and RFP form of specifications is the omission of the product make and model number.

Determining what products to specify is a three-step process. The first step is to list all the design criteria and design details of the system as determined at the assessment and design phases of the project. In other words, what are the needs of the system? This is all the work that has already been done. The second step is to make a list of products that meet those needs. This is crucial: MATCH PRODUCT TO NEED, NOT NEED TO PRODUCT. The third step is to choose the BEST products and equipment for the system.

Before this process can take place, the consultant must have an understanding of the products available in the industry and how they function and operate. This is one of the reasons why a solid technical security background and experience in the design and implementation of physical security countermeasures are prerequisites to becoming a security design consultant. However, in today's world, security product technology is changing constantly, seemingly every day. It is imperative that security design consultants keep up with the changes so that they understand what is available in the marketplace to fit the needs of their clients. In addition, they need to know the details of incorporating the available products into a correctly functioning system.

HOW TO OBTAIN PRODUCT AND TECHNICAL KNOWLEDGE

The security design consultant must continually obtain knowledge of the products available in the marketplace and the new technological advancements regarding how those products function and are incorporated into a working system. A recent example would be with the advent of IP cameras. Security design consultants, and contractors for that matter, not only had to know that IP cameras were an option in the marketplace, but had to learn about IP technology as it applies to the cameras to know how they are wired for both power and video signal and how they are combined with other products to form a functioning system. Additionally, products enter the marketplace all the time that don't necessarily utilize a new technology but have better or different features than what already exists or have an improvement on the existing features of the same product (DVR hard drive size as an example). The following sections describe ways in which the security design consultant can continually be educated about these new developments so when the time comes to design a system, the consultant has all the available product options to choose from.

Conferences and Expositions

At large annual conferences and expositions, hundreds of manufacturers gather together to demonstrate their existing and new products and answer attendees' questions. The manufacturers may also give short learning presentations about their products. The conferences themselves include educational sessions on a variety of subjects and new technologies. Most industry professionals are already familiar with these conferences; they include the American Society for Industrial Security (ASIS) International Annual Seminar and Exposition, the International Security Conference (ISC) Expo sponsored by the Security Industry Association (SIA), and the National Fire Protection Association (NFPA) World Safety Conference and Exposition. The contact information for these organizations is included in Chapter 13. The literature, CDs, and other information gathered at these events become a resource library for the design consultant that can be referenced when a project need has to be filled. All security design consultants have space in their offices dedicated to the storage of these materials.

Working with Manufacturers

Going directly to the product source—the manufacturers—is an excellent way to gain knowledge. It is not necessary to wait for the next conference to see what's new; almost all manufacturers have websites where all their product literature is available to download. User manuals, specifications sheets, CAD drawings, software, A&E Specifications, and sales literature may all be available to some degree. Also, most manufacturers have a technical support telephone number where technicians are available to help with any design questions. This is a well-used resource for security design consultants and most of the time well worth the on-hold wait commonly experienced. If the consultant doesn't know a manufacturer for a certain product type, a Google search or review of one of the available security industry buyers' guides will find what is needed.

Some manufacturers have special consultant programs to support security design consultants and help them understand their technology and specify their products. This is a relatively new concept within the industry but has turned out to be very successful both for consultants and manufacturers. Employees of the manufacturer assist security design consultants with issues of specifications, literature, technical support, pricing, and contractor locating. Many times, these people are former security design consultants themselves. Manufacturers with these consultant programs will sometimes hold seminars that last a day and teach about a specific technology or manufacturer product. If security design consultants feel they need assistance in understanding a particular technology, many times these short seminars are very helpful.

Invariably, some manufacturers are better than others at the availability of their literature, the quality of their technical support, and their overall support of security design consultants. Some manufacturers won't even talk on the phone with anyone who isn't an authorized dealer of their equipment. They just don't understand that there is a whole group of competent people out there who may be willing to specify their products. Other manufacturers understand that point very well and go to great efforts to be helpful. Security design consultants should work with those manufacturers who are willing to help; plenty of them are out there.

To reiterate a point made earlier, a professional and ethical design consultant should never specify a particular manufacturer or product just because he likes the company, or company representative, or because that person was helpful. The issue is, all else being equal, if the manufacturers are willing to help before any guaranty of a sale, the odds of them being helpful during and after the sale are better, which is better for the project and the client in the end.

Working with Manufacturers' Representative Firms

Many manufacturers market and sometimes sell their products through manufacturers' representative firms. A manufacturer's representative firm is a company that represents many manufacturers, usually 10–20, within a specific industry. Some firms will represent any company within the security field; others will specialize in a manufacturer of a certain type of equipment, CCTV equipment as an example. The job of the manufacturer's representative firm is to educate the contractors, distributors, consultants, and end users within its geographical area on the merits of the manufacturer's equipment so that equipment is ultimately sold and installed. The geographical area covered by one firm is usually regional, encompassing many states, and the manufacturer uses multiple firms to cover the country. Some firms also are allowed to take sales orders for the products; in either case the sales of the products within the firm's region are credited to the firm and the firm is compensated accordingly. When a consultant contacts a manufacturer for information, someone will tell the consultant if there is a manufacturer's representative firm used and who it is for the appropriate region.

Because of their function, manufacturers' representative firms are, for the most part, very willing to help security design consultants. They understand that if the product is specified within the design specifications, especially for an IFB, the project will include those products and some contractor will be purchasing the product. They can help in similar ways to the manufacturer itself or a consultant contact within a manufacturer, with specifications, CAD drawings, technical support, overall technical questions, pricing, and other pertinent issues. In many cases these firms also have demonstration equipment available and will demonstrate for the consultant the functions of the product. This is generally very educational and a great help to the consultant. Some representative firms may be hesitant if the project is outside their geographical area even though the consultant is within the area, but once a relationship is established with the firm, even location usually isn't a problem.

Working with Distributors

Probably the most common way security products are sold is through distributors. There are national distributors such as ADI, as well as many smaller, regional distributors. The problem with distributors from a security design consultant's point of view is that their sole purpose is to sell any product, and unlike manufacturers who will make the sale regardless of who purchases the product, distributors have no guaranty of the sale because the consultant is not the one buying. Therefore, they have little incentive to help with technical information or their most helpful attribute for the design consultant, pricing. If the security design consultant comes from an integrator background, the relationships already developed with a distributor will perhaps help overcome this situation, and the consultant can get some assistance when needed. Distributors also provide catalogs of the products they sell with some pricing information that may be helpful, and occasionally they host some manufacturer seminars or small, one-day expositions to help fill in the gap between the larger ones.

Publications

Numerous publications in the industry advertise and promote new products from manufacturers. They usually come out monthly and describe new products and technical applications that may help educate the consultant. The volume of these publications can be overwhelming, so the consultant should subscribe to just a few of them that he feels are the best and diligently scan them for information.

Now that the security design consultant has the knowledge of what products and technologies are available and can find the detailed information that may be required, the first step in the process of determining what products to specify can proceed.

SYSTEM NEEDS: DESIGN CRITERIA AND DETAILS

The first step in the process of determining what products to specify is to organize all the design criteria and details. The design criteria and design details of the system were established during the assessment and design phases of the project. The only thing left to do is organize the information in some logical order to make it easier to match a product to these needs. The information should be grouped by system, including the elements of all necessary products for each system so they can be viewed in their entirety. Other than that, how the information is formatted is a matter of personal preference. The important point is that it is all written down in one place and organized in a manner that the consultant can understand so that no detail is missed. It would be a shame to choose the wrong piece of equipment not because the effort wasn't put into getting the design details, but because something was forgotten in the selection process.

MATCHING PRODUCT TO NEED

The second step in the process of determining what products to specify is to determine what products meet the needs of the design criteria and details. The most important concept in this chapter is that the job of the security design consultant when it comes time to choose products to specify is to MATCH PRODUCT TO NEED, NOT NEED TO PRODUCT. In other words, the consultant needs to pick the products to make up the system that meet all the design criteria and design details as established during the project. He should not pick a product or manufacturer of products first or try to find the one that fits the needs the closest. This approach may seem like common sense, but it is not practiced in the industry anywhere near as often as it should be, and the result is lousy physical security countermeasure systems that don't accomplish their intended purpose.

Matching product to need is one of the selling points for a client to hire a security design consultant to design a system instead of going directly to an integrator. Integration or installation companies have a tendency to represent a small number of manufacturers most of the time. This is due to their having a comfort level with those manufacturers, volume sales quotas that need to be met, incentives from manufacturers, or sometimes it's just easier. Maybe the manufacturer chosen has a piece of equipment that exactly meets the needs, and maybe it doesn't. Even worse are the large companies that manufacture, sell, and install their own equipment. Their product selection is even more limited than the integrators' selection. A security design consultant should not be limited in the number of products or manufacturers available for consideration.

There are, however, other traps that designers can easily fall into when choosing manufacturers of equipment. Traditionally, these discretions have been committed by architects, but with the growing popularity and acceptance of security design consultants, it has become common for unprofessional consultants to be guilty of the following as well:

1. Some manufacturers have a tendency to wine and dine system designers with the goal of having those designers specify their equipment almost exclusively. A well-known fire system manufacturer (whose name shall be omitted) is notorious for this behavior, although architects are its main target. While this behavior is not necessarily unethical on this manufacturer's part, it certainly is unethical on the designers' part. Now, it is quite common for manufacturers to take consultants to lunch or other such thing and explain their product lines. This is certainly not unethical behavior on the consultants' part; in fact, it is a good thing to do to help keep up with the products available to them as well as develop strong and lasting relationships with the manufacturers that will help the consultants in the long run. It's when there is a quid pro quo that there is a problem.
2. Additionally, some manufacturers will offer to write the specifications for designers, of course, specifying their equipment, at no charge. A well-known burglar alarm manufacturer and installation company (again whose name shall be omitted) does this as part of its normal business operations. This is definitely unethical on the designers' part, as they are charging for work performed by others, in addition to the end result probably not being the best choice for the client.
3. As with integrators, designers can sometimes develop a comfort level with certain manufacturers and specify their equipment regardless of their clients' needs. These designers may not have the financial obligation as the integrators do with quotas, but may believe they have a loyalty obligation. The result is not providing the best option for the clients. That is not to say that over the years security design consultants will not or should not develop a list of preferred manufacturers that they have had success with and start with these manufacturers when looking for products. That is only human nature. As long as the consultants will look elsewhere if the needs of the project are not met by any of those manufacturers, there is no problem.

It is not just which manufacturer is chosen that is important, but which products from that manufacturer are chosen. When the products chosen do not meet the needs of the client, the security design consultant failed to use best practices for the project and ultimately failed to do the job for the client. The errors may be egregious or subtle, and the client may know about them immediately or not know about them until some time after the project completion. Following are some examples of products not meeting the needs of the client:

1. A camera system is designed for a police station in a small town. The technical aspects of the design are perfect, and the system includes pan-tilt-zoom (PTZ) cameras, fixed cameras, monitors, digital recorders with network connection capability, and PTZ head-end equipment. All the equipment specified is of excellent quality, and the system works perfectly. However, the police department is small, including only 11 officers and one part-time secretary. Most of the time no one is actually in the building. No one is there to use the PTZ feature of some of the cameras. Having them as part of the system is a colossal waste of money; instead, having multiple fixed cameras to see more of an area than the one PTZ camera could see, or a simple automatic swiveling mount for the single camera would be more appropriate and effective. Errors in meeting needs can include not having enough equipment or features or having too much equipment or features.
2. A hydraulic slide gate operator system is designed for a self-storage company with facilities all over the country. The systems are installed in several facilities and work to perfection. Then one of the facilities is constructed in the northeast part of the country, and the same system is designed and installed. When the first ice storm hits, the ice builds up on the drive rail, and the rollers no longer move the rail through the operator. The system becomes useless and inoperable during many parts of the winter. The designer failed to take into account the environmental conditions of the area and therefore design the appropriate chain-driven slide gate operator system for that facility.
3. A fire system is designed for a building on a college campus. The same manufacturer of the equipment already installed in the rest of the campus buildings is used. However, the new model fire panel style is used for the design instead of the same panel series as used for the rest of the buildings. The system works well for a couple of years. Then the college decides to operate its own monitoring facility that will encompass all its security and fire system monitoring. The software used for the new fire panel model is not compatible with the software for the rest of the fire systems. The designer did not check the compatibility or anticipate future expansion.

To have the product meet the needs of the facility and client, a set of criteria should be used to make that determination. Every potential product should be vetted through the criteria until a few products that meet all the criteria are found. This will be easy for some applications and not so easy for others. Many times in an assessment report, the following criteria are listed as the reasons for choosing the products on the short list:

1. *Function:* Do the products perform all the functions as determined by the design criteria and design details? This is the most obvious criterion for choosing a product. Every single design criterion and design detail must be accounted for and match the product. Incorrect functional performance can have either too few or too many functions than required to meet the design needs. The consultant also must make sure that ALL the required products are specified to create a complete working system. The size of the power supply is just as important as the functions of the head-end equipment.
2. *Compatibility:* Are the products specified compatible with each other and able to work in concert to create that complete working system—not just in general function but with compatible wire connections, software protocol, mounting, and so on? Does the system have to work in concert with other systems in the facility? If so, are the products compatible with the existing equipment?
3. *Ease of Installation:* Are the installation techniques required to properly install the products the normal industry standard installation techniques, or do the products require special training or a skill that would either greatly increase the cost of the system or severely limit the installation company options? If something special is required, it doesn't automatically eliminate a product, but this consideration must be taken into account when evaluating all the potential products. Also, some products are just easier to install than others. Knowing this information takes either personal experience or being told that by integrators who have installed those products in the past. The consultant should ask them at the end of a project how easy or hard installing the specified products was. Integrators are a great resource.
4. *Availability:* The availability of a product is important. If it's hard to find or purchase the product, keeping an installation schedule will be difficult. If it's hard to get information on the product from the start, it probably will be hard to purchase.
5. *User-Friendliness:* Is the operation of the system made from all the products user-friendly for the client? Is it convenient, aestheti-

cally pleasing, and easy to understand? If a system's operation is too difficult, it either won't be used or not used to its full capability. The consultant should imagine he is the client and walk through the operation steps himself.

6. *Expandability:* Are the products or complete system made up of the products expandable in order to anticipate potential future requirements? There is a fine line here between expandability and having more functions than are needed. An overly simple example is specifying a 16-camera capacity DVR instead of a 4-camera capacity for a 4-camera system. One of the interview questions during the assessment phase of the project was about potential future requirements. The consultant should use this information to make the best judgment based on best practices and experience.
7. *Reliability:* Is the technology represented by the product reliable and proven? The consultant does not want his client to be a guinea pig for something unproven that turns out to be unreliable. Is this the manufacturer's first model of this type of product, or has it made this type of product for some time so the bugs have been worked out? Is there any track record that can be referenced for this product? If possible, talking to clients who have the product installed or integrators who have installed the product is an excellent way to make this determination.
8. *Client- or Project-Specific Criteria:* Sometimes the individual client and/or project will have specific criteria for a system that is not universal. These criteria will have been determined in the assessment phase of the project. They could be anything from the products having to be UL listed to products that must be purchased through a specific channel. Some clients have some strange criteria, but as long as the resulting system meets the needs of the client, that's okay.

CHOOSING THE BEST PRODUCTS

The third step in the process of determining what products to specify is to determine the best products available for the needs of the system. Now that the consultant has a short list of some of the products that meet all the criteria and the system's needs, he must choose the best products. Sometimes in an assessment report, this list is written along with the reasoning for choosing the best product. This process is both objective and subjective, both in the specific criteria and the weight given to each criterion. There is no set rule of thumb here; it is up to the consultant's expertise and best practices. The consultant must be able to explain the reasoning, however. The criteria used to choose the best product include following:

1. *Price:* This criterion is simple: which product that meets the needs is cheaper? There is no sense in paying more than necessary for the products that make up the system. Just because a product is more expensive does not mean it is better. Some less expensive products are of inferior quality, but those should have been dismissed using the previous product criteria.
2. *Manufacturer Comfort Level:* What is the consultant's comfort level with the manufacturer? Does the manufacturer have a good reputation in the industry? Is it easy to deal with, and does it provide good service before and after the fact? This is where the consultant's personal preferences come into play, where the relationship with the manufacturer or its representatives influences the decision. There is nothing wrong with that, since it's already been determined that the products meet the needs of the client.
3. *Installation Pool:* Is any qualified contractor able to purchase and install the products? Some manufacturers have only specific dealers or integrators, sometimes called *business partners,* who are allowed to purchase and install their equipment. Most of the time there are only a few of these dealers in a specific geographical area. That may or may not be a problem for this specific project and specific client. However, situations in which only one company is allowed to install a system should be avoided because they eliminate the client's ability to solicit a competitive bid.

These criteria must be given weight and evaluated, and a decision is then made as to the best products to make up the designed system. The consultant is now ready to write any form of design specifications required for the project.

9

Determining Total System Cost

Part of the analysis in the preceding chapter was comparing the costs of potential physical security countermeasure product choices to help determine which product to specify. It was not necessary for that analysis to use every component of a system cost, just the ones that vary from product to product to make a comparison. Once the product choices have been determined, it is necessary to quantify an estimated cost for the implementation of that system. This is a very important service to offer the client.

In many instances a preliminary determination of the total system cost needs to be made during the assessment phase of the project, whether or not the exact products are chosen as described in Chapter 6. In that case, the methods of making that determination as described in this chapter are all the same, except the potential products or set of products described in Chapter 8, which are used for the analysis based on the results of the assessment and the expertise of the consultant. The results of that determination will be used as the system budget, with a recommended 20–25% contingency, until this analysis is done again after final product determination.

Whether or not a preliminary cost determination is made during the assessment phase of the project, the budget for the system cost needs to be established before the procurement phase of the project can begin. No one really wants to go through the trouble of a bid process without knowing whether he can afford it. This budget estimate should be done whether the procurement method for the project is an IFB or an RFP. If an IFB is utilized, the products specified are used for the analysis. If an RFP is utilized, a set of potential products based on the system requirements and the consultant's experience should be used for the analysis. This estimate may be the same as one done during the assessment phase or different if new information has come to light or changes are made during the design phase of the project. However, this estimate should now have a contingency of only around 10–15%.

If the consultant is working with or for an architect on the project, there may be additional phases of the budget process. The architect may want budget estimates during the assessment (planning) phase, at the 50% of design completion phase, and at the end of the design phase. This can be the case for very large projects where the design is a long fluid process in which decisions change as the design moves forward.

Determining total system cost (sometimes called *life-cycle cost*) is not as simple as getting a proposal from an integrator for the system installation. While the cost received will be for the entire installation, costs other than the installation costs make up the total system cost. Taking into account every cost the client will realize from the implementation of the particular physical security countermeasure is another factor that separates the services of the design consultant from the client just calling the nearest installation company from the phone book to plan a budget for the project. The following are the cost components that must be taken into account:

1. System Design Costs
2. System Installation Costs
3. System Operation Costs
4. Maintenance Costs
5. Replacement Costs

SYSTEM DESIGN COSTS

There is a cost associated with the design of the system, including specifications, drawings, and other procurement documents. Hopefully that cost is the fee of the security design consultant working on the project. However, even if a design consultant is not used on the project, someone has to do that work, and that cost is part of the overall system cost. The fact that this cost exists is why some potential clients will call an integrator, tell him what they want, have the integrator design the system free, and then install it. That is why the consultant will in many cases need to justify the quality and necessity of his services.

SYSTEM INSTALLATION COSTS

System installation costs are traditionally considered the costs of the system, what the response would be to an IFB, or what an integrator or other installation company would charge to install the designed system. To estimate this cost, one must think like an integrator and take into account all of the costs of installation. The components of installation cost include the items described in the following sections.

Product Costs

Product costs are what the equipment costs are to the integrator. Just as in obtaining general product knowledge, it is important to have a relationship with distributors, representative firms, and manufacturers to obtain these product costs. Using the suggested retail price for these product costs is not an acceptable way to come up with the equipment cost. Every manufacturer has a different percentage increase to come up with its suggested retail price. These increases are generally anywhere from 10–50%. Therefore, using these retail prices, especially if there are products from different manufacturers within the same system, would skew the price too much for a proper estimate. It must be determined within a reasonable percentage what the installation company will pay for the products. Determining these prices will take a little extra time, but it is a necessity.

Shipping Costs

The equipment has to get to the integrator and/or the end user site. Where is the equipment coming from, and how much will shipping cost? This amount may seem almost irrelevant in the big picture, but in fact shipping costs on larger items can be quite expensive. As with the cost of the products, manufacturers, distributors, and representative firms can help in making these determinations.

Labor Costs

The cost of labor is usually a major part of the system installation cost and has many variables. There are two elements of labor costs that need to be taken into consideration: the labor rate and how long the installation will take.

To determine the first element, the labor rate, the consultant will need to do some research to answer the following questions:

1. Exactly what type of trade will be required for the installation of this particular system? For instance, does it require a security integrator, electrician, gate operator company, or any combination of trades for multiple facets of the installation? Anyone with technical security experience should have a good understanding of these answers.
2. What is the typical wage rate for that trade or trades in the geographical area of the project? The same job costs much more in labor in Boston versus Fargo. A few phone calls should garner this information.
3. Is prevailing wage required on this particular project? In many government projects and even some private ones, the wage for each classification of worker is determined by a prearranged wage scale, usually by the state. Because these wages are in most cases considerably higher than a trade contractor would normally pay (government work doesn't cost a lot for no reason), the consultant needs to ask this question early in the process. If it is a prevailing wage job, those rates can be easily found on the Internet and question #2 is moot.
4. What are the different rates for the different job functions within the project? For prevailing wages, this is definitely the case. For any good-sized integrator or electrician, this is also true. The different job functions are described in the following list. Again, the consultant will need to do some research to make those determinations on a project-by-project basis.

When the consultant is doing research, he must keep in mind that some of the hourly wages (nonprevailing wages) may be what the contractor charges hourly for the service rather than what the employee is paid. If this is the case, then the profit/overhead numbers described later need to be adjusted downward to reflect that.

Making a determination on the second element of labor costs, how long the installation will take, requires industry expertise. This is another one of the reasons that only individuals with technical security experience doing other jobs should become security design consultants. Some experience is definitely required to estimate this time. The time can be broken down into the following job functions:

1. *Site Installation:* This is the time it will take the technicians on site to wire and install all the required equipment. This part may also be broken down into different levels of technicians, such as a senior technician versus a wire puller.
2. *In-Field Supervision:* Most projects of any reasonable size use a site supervisor or project manager to oversee the work on site. Without this person to keep the job on track, the installation company's labor costs can get out of hand quickly. This person usually has a higher labor rate than technicians.
3. *Programming/Engineering:* There is usually an in-house person or persons doing the AutoCAD work for shop drawings and as-built drawings, as well as programming system software on the PC. The particular job determines how much this person is needed, and again, his rate is usually different.
4. *Testing/Training:* Will these functions be done by a supervisor or technician? The answer is determined both by the size and type of project and how much of each is required. The design consultant, in many cases, dictates through the specifications how much time these functions will take, so it's just a matter of who will be doing the work.

As stated earlier, there is no substitute for experience when estimating these times. However, if a consultant does not have that experience or needs a refresher, spending time with an integrator in the field can help. If the consultant does not have an existing relationship with an integrator, a relationship should be developed. Spending some time with installers on a job site and a programmer in the office will help immensely in budget estimation. It will help even more for project management, as described in Chapter 11. If there is a concern about ethics when developing relationships with integrators, the consultant should be assured there is no ethical wrongdoing in ensuring an integrator is put on the bid list for projects he is qualified for. The consultant should make sure there is not the perception the integrator is guaranteed the work. Most integrators will be quite happy with that agreement. If that's not the case, the consultant should move on to one who is.

Fixed Costs

There are many fixed costs, for lack of a better term, associated with a project that must be taken into account. They can include the following:

1. *Material Costs:* Materials include hardware, screws, tape, wire ties, labels, paint, etc. Usually, there is a fixed cost put into a bid price for these costs, depending on the size of the job.
2. *Subcontractor Costs:* In some cases the labor figured previously will not include all the work required for the job. For instance, if an integrator is installing a CCTV system but must also provide the power to the power supply, he may not be licensed for that work. The integrator may have to hire an electrician to provide that power, and that cost needs to be accounted for.
3. *Bonding Fees:* Many government jobs and larger private jobs require bonds. They come in two forms: bid bonds and performance bonds. A *bid bond* is put in place to verify the bidder will honor the bid, and a *performance bond* is put in place to verify the work will be done to the client's satisfaction. A bonding company provides the bond and charges the bidder a percentage of the total job to provide the service. It needs to be determined early on if bonding is required, and that cost needs to be determined.
4. *Permit Fees:* In many cases the local municipality requires a fee for the permit to install a new system. The fee is usually based on a price per thousand dollars of system cost. The local building official's office will have the fee schedule and an understanding of what does or does not require a permit.
5. *Taxes:* Depending on the type of client and geographic area, state and local sales taxes may apply and need to be figured into the cost of the system.
6. *Tools:* Is there a specialty tool needed for this project that needs to be paid for? Special lifts, saw blades, programmers, or testers are some examples.

Profit/Overhead Costs

Up to now all the costs have been what the bidding company needs to spend to complete the installation. Now the company needs to pay the light bill and actually make a profit. The following can be included in this category:

1. *Sales Commission:* If there is a commissioned salesman bidding on the job for the company, that fee needs to be added. If there is no commissioned salesman, there is still some percentage fee associated with the man-hours to bid the project.

2. *Overhead:* This is to pay for the rent, lights, etc.
3. *Profit:* The company needs to make money to be viable and continue to stay in business.

The reason these items are lumped into one category is that there is usually some percentage *markup* used that is inclusive of all these items. For instance:

Total product, shipping, labor, and fixed costs	$76,000
Markup	× 1.4
Total cost of installation	$106,400

This percentage varies from company to company and situation to situation. Because of competitive bidding, there is no steadfast rule here. Knowledge of the marketplace within the geographical area of the project and within the type of company bidding is necessary to make an accurate estimation of the markup. This variable, along with the variables in the other categories, makes it virtually impossible to estimate the cost of installation exactly. However, the consultant is not being asked to be exact but to be close for budget purposes. Being within 10% of the actual bid prices is considered good and within 5% is considered excellent. Experienced consultants take great pride in being as close as possible to the actual bidding price.

Some companies, especially larger ones where the costs are determined from a central office, will have set product prices and labor rates per job function that include the markup. The preceding method should yield the same result.

Table 9.1 shows a sample spreadsheet that can be used to estimate the total job installation cost. This type of spreadsheet can be created in Microsoft Excel or similar software and can be customized to the user's liking. The software allows for formulas to be assigned to specific fields so all the math is automatically done after the numbers are plugged in. Table 9.2 shows the same spreadsheet with a sample job cost estimation. The categories under each main heading can be as general or specific as needed to accurately do the estimation; the sample depicted is fairly general. If this were the spreadsheet actually used by the installation company for bidding purposes, it would more than likely be much more specific, including all actual product quantities, makes, and prices.

Table 9.1 Sample Blank Installation Estimate Spreadsheet

INSTALLATION ESTIMATION FORM			
TASK			**PRICE**
Total Product Cost			
Shipping Cost			
Labor Cost	**# of hours**	**Rate**	
Fixed Costs			
Sub-Total			
Markup			
Total Installation Cost			

Table 9.2 Sample Job Cost Estimation

INSTALLATION ESTIMATION FORM			
TASK			**PRICE**
Total Product Cost			
Control Panels			16,350
Power Supplies			1,230
Field Devices			10,960
Software			890
CPU			1,810
Shipping Cost			**170**
Labor Cost	**# of hours**	**Rate**	
Site Installation	160	40	6,400
Supervision	40	50	2,000
Engineering	25	60	1,500
Training	10	40	400
Fixed Costs			
Material			500
Subcontractor			1,500
Permit			1,000
Sub-Total			**44,710**
Markup			**1.4**
Total Installation Cost			**62,594**

SYSTEM OPERATION COSTS

Installed physical security countermeasures do not work in a vacuum. They all require some degree of interaction with personnel, either directly or indirectly, for them to function. This interaction has an associated cost that would not exist except for the installed countermeasure. That cost needs to be taken into consideration when evaluating the total cost of the system. For instance:

1. Do personnel have to be added to operate the system? Does a new guard or central station operator need to be hired to accomplish this task? Does this task take time away from existing personnel so that their existing workload needs to be made up somewhere else?
2. Does any new infrastructure have to be installed and/or maintained because of the system? For instance, is a new guard desk or separate section in the central station required?
3. Is supervision or administration required to have the system function properly? For instance, does someone need to run and analyze access control reports on a regular basis or view CCTV images regularly?
4. Does there need to be interaction with other departments, such as the IT department, to ensure the system works smoothly in its interaction with the rest of the company or facility?
5. Do existing policies and procedures need to be updated and regularly reviewed as a result of the newly installed system?

These costs all need to be estimated and added to the total system cost for analysis. The client will probably have to be involved a great deal in determining how much of this type of cost applies and what the cost actually will be. The consultant's job is to facilitate in helping the client to understand these costs need to be accounted for, finding out what they are, and adding them to the total cost analysis.

MAINTENANCE COSTS

Physical security countermeasures don't work in a vacuum; likewise, they don't work correctly all the time, no matter how well they were designed and installed. Just like the building itself and the copy machine, these systems need to be maintained to keep downtime to a minimum and to ensure the longest life possible for all the equipment. For instance, if a gate operator does not receive maintenance on a regular basis, in no time it will break down and cost more to repair than the cost of the maintenance or, even worse, to have to replace it. The following questions need to be asked:

1. What routine maintenance must be performed on the equipment? What does the manufacturer recommend? Does the environment in which the equipment is installed affect it at all?
2. How often does this maintenance need to be performed?
3. Who needs to perform this maintenance? Does the installation company need to perform this service, or can the end user's maintenance personnel be effectively trained to perform the functions?
4. Even with maintenance, how often is the equipment expected to be nonfunctional and need repair? Are additional personnel needed when the equipment is nonfunctional?
5. What are all the costs associated with this maintenance and repair?

This cost as well needs to be estimated and added to the total system cost for analysis. One option to take care of this issue is for the client to purchase a service or maintenance contract from the installation company. Depending on the details of the individual contract, all routine maintenance and replacement parts are taken care of with a fixed contract price, regardless of what goes wrong.

REPLACEMENT COSTS

Once the end user is committed to the installation of the physical security countermeasure, that user is generally committed to it long term, or at least probably should be. That means as different pieces of equipment break down and it is no longer cost effective to attempt repairs, they need to be replaced. A good client should not wait for, nor should a good consultant recommend to, the inevitable before planning financially for replacement. Two questions should be asked:

1. What is the life cycle of all the equipment within the system? This information for some equipment can be gathered from the manufacturer. For other equipment, industry standards and the consultant's expertise are required.
2. At the time the equipment needs to be replaced, how much will it cost? This amount can be reasonably estimated based on what it would cost today plus a standard increase for the number of years until replacement.

The total replacement cost needs to be spread out over the total number of years until replacement to analyze total system cost. Keep in mind that the equipment that will ultimately replace this existing equipment may be very different due to technology upgrades or the different needs of the client, but at least a reasonable dollar figure is being budgeted. Another option is for the client to purchase a long-term service contract in which the installation company replaces all defective equipment as part of a fixed contract price.

TOTAL SYSTEM COST

Following is an example of total system cost for the first year and subsequent years:

System Design Costs	$ 18,000
System Installation Cost	$154,000
System Operation Costs	$ 14,000
Maintenance Costs	$ 4,000
Replacement Costs ($175,000 ÷ 10 years)	$ 17,500
Total Yearly System Cost: First Year	$207,500
System Operation Costs	$ 14,000
Maintenance Costs	$ 4,000
Replacement Costs ($175,000 ÷ 10 years)	$ 17,500
Total Yearly System Cost: Subsequent Years	$ 35,500

Some clients may not want to figure the future replacement cost into the yearly ongoing costs of the installed system, but the consultant should always at least offer the information to the clients.

COST/BENEFIT ANALYSIS

After the consultant determines the total system cost, there should be some analysis as to whether protecting the asset is worth the cost. The consultant needs to work with the client to make these decisions, and they can be a topic of discussion after the client receives the assessment report. Each of the three general asset categories should be taken into consideration in this process:

1. *Hard Assets:* The client would not spend $10,000 to protect a cash register with $100 in it. However, the client probably would spend $10,000 to protect a safe with $1,000,000 of jewelry

inventory. If the asset is the building itself and all the contents, the same analysis applies.

2. *People:* Is $10,000 worth protecting employees in a parking lot? Even if the client doesn't care about the employees (to be pessimistic), is protecting them worth not having the lawsuits when an incident happens in the parking lot?
3. *Information:* Is $10,000 worth protecting the formula for the client's signature product?

The answer to the cost/benefit decision may be based solely on the perceived needs of the client. However, this basic exercise, which consumes little time, should be done just to make sure the decision is a sound one.

BUDGET DISCREPANCIES

Discrepancies between the estimated system budget determined by the consultant and what the client can or is willing to pay for the system can come at different points in the project process. The first may be after the consultant does a preliminary total cost estimate during the assessment phase of the project and, whether or not the cost/benefit analysis is favorable, the client is not willing to spend that amount of money on the security countermeasure. This discrepancy is the easiest to fix because it is early in the process and is a big reason why a total system cost estimate should be part of the assessment process. The consultant should work with the client to find a type of system that will fit into the client's budget while still protecting the asset. The possibility exists that the client is simply not willing to spend the required money to properly protect the asset. In that case, ethics dictate the consultant should not design a system that clearly will not protect the asset.

If there is no assessment report phase of the project but simply the design of a system with a fixed budget, a discrepancy may exist between the total system cost of what the consultant designs and the fixed budget. There is no one answer to fix this problem. The following questions should be asked:

1. When the choice of products was made, were there products that may not have been the best for the application but were on the short list and were less expensive enough to make a difference? The consultant must keep in mind that the needs of the client must still be met by the products.
2. Can the system installation be phased as to allow for some of the system to be installed now within the budget and the rest installed

at a later time? How does this approach affect the operational costs?

3. Is it reasonable that the cost of the type and size system being designed be close to the budgeted amount?

The consultant may need to inform the client that the budget is unreasonable based on the type and size of the system being designed. In that case, the client has to decide whether to adjust the budget, phase the project, or cancel the project altogether. Every consultant will eventually have a project canceled because the client simply did not budget enough money.

One important consideration the consultant must keep in mind is never to grossly underestimate the estimated cost of the system for any reason, even a virtuous one. Not much is a worse hit to the credibility of a design consultant than to have the actual bids of a system come in considerably higher (more than 20%) than he estimated. This outcome is fairly indefensible. If the bids are considerably lower, the client might not get upset, but a good consultant doesn't want this to happen either. If the process described in this chapter is followed and the consultant does the proper research to garner good information, a reliable budget estimate for total system cost will be obtained.

10

The Report

There are two main types of report writing for security design consultants. The first is the assessment report, which can include either the results or findings of a physical security assessment or the elements of system design, with or without product identification, or both. The second is the actual writing of design specifications and their associated drawings that are ready to go out to bid along with the client's bid documents. Many projects include an assessment report, but then the consultant never takes the next step of writing design specifications. However, if design specifications are written, in most cases an assessment report has been written with subsequent approval of the report's conclusions by the client prior to the specifications being prepared.

THE ASSESSMENT REPORT

The assessment report is the way the consultant conveys his knowledge and recommendations to the client. There are many potential elements or sections to an assessment report. Which elements are used in the report depend on the scope of services, type of client, size of project, and personal writing style of the consultant. Appendix C shows two sample assessment reports. The first is from a security design consulting firm, and the second is from a security management consulting firm where the design consultant partnered with the management consultant to perform the physical security aspect of the assessment.

Assessment reports should always be in writing and not just delivered orally. There is too much that can be misinterpreted or misheard when a report is just given orally. In addition, there needs to be a record of the consultant's work for the benefit of both the consultant and client. It is appropriate and common for a consultant to give an oral presentation to a client in addition to providing the written report. In most of these cases, there is some kind of presentation to a committee or board. Before preparing for such a presentation, the consultant should find out from the client how many people will be there for the presentation, who they are, and what is expected regarding the presentation format. That way, the consultant can tailor the content of the presentation to the audience and not be caught off guard as to its expected form. In some cases the consultant just sits at a table with the group, reviews the report, and answers any questions the group may have. In other cases the consultant is expected to give a formal presentation that may also include a PowerPoint presentation. In any of these cases, the consultant should keep the presentation simple and straightforward and take the audience through the same logical process he went through to reach the conclusions.

Depending on the client or project, providing a draft copy of the report to the client for review before preparing the final report and submitting it may be appropriate. Draft copies are not as accessible to everyone from a legal standpoint as the final copy, and the client may have good reason to ask the consultant to leave something out of the final report. As long as the removed item does not change the conclusions arrived at by the consultant, there should not be any ethical issue regarding removing some item or language at the client's request. This issue should be discussed with the client before the beginning of the project.

Following are descriptions of many of the sections that can be included within an assessment report. They may be able to stand alone as their own sections or may be combined with each other or incorporated into other sections, depending on the project and the individual consultant's writing style. Not all of these sections will be used in every report, and some report sections appropriate for a project may not be listed here.

Executive Summary

An executive summary should be included in all reports. The executive summary is exactly what it says, a summary of what the report is all about. The executive summary should include the following information:

- A brief synopsis of the project
- A brief description of the consultant's qualifications and company information
- A description of what has been done during the project by the consultant up to the writing of the report
- The categories of information included in the report
- A list of any attachments to the report

The executive summary is important because not everyone reading the report will be familiar with what has taken place so far in the project, or all the details of the project itself for that matter. After reading the executive summary, the readers should know exactly what they are reading about and what information they will receive from reading the whole report.

Summary of Recommendations

In some cases, a list of the recommendations without full explanations of them should be included after the executive summary. Most of the time, this summary would be included when there are a lot of recommendations throughout the report, so an individual recommendation might be hard to find.

Elements of Design

Elements of design should be included in reports in which the scope of the project is not just to recommend a system that should be installed, but the what, how, why, and where of that recommended installed system. In this section, all the design details are described. This description should be very detailed, including how the system will operate and function; the exact locations, sizes, and types of all devices and head-end equipment; pertinent specific information about devices and head-end equipment; and any other necessary design elements. Anyone with knowledge of the facility should be able to read this section and follow along, understanding where everything is going and why and how it will function. Although some conceptual drawings of specific equipment may be included with this report, there probably will not be drawings of the facility with device locations marked, so the reader has only the written words of the report to paint a picture of the system.

Product Selection

The report should include product selection if it was part of the scope of the project and writing design specifications and drawings for an IFB will be the next step in the project. This section has three distinct parts, which mirror the logical process described in Chapter 8 for choosing the correct products for the needs of the facility. The first part is a listing of all the criteria that were used to choose equipment. In other words, what are the requirements or needs of the system and products within the system based on the design criteria and design details? The second part is to provide a list of a few products or systems that meet those requirements and needs. The product is being matched to the need, not the need to the product. The third part is to recommend one of those products or systems and to explain the reasons for the chosen products. This section is a logical thought process from what is needed and why to what is chosen for that need and why. If this section is done correctly, the client should have no issue with the selected product or system.

Estimated Costs

The report needs to include an estimated cost for the system being recommended if system design is part of the report, whether or not exact products were chosen and included in the report. The results of the analysis, as described in Chapter 9, need to be provided. If multiple systems are being recommended, then it may be appropriate to provide an estimated cost for each of those systems. The other items that may be included in this section are the budget for the project, how the estimated costs fit or don't fit into that budget along with an explanation if they don't, and a recommendation as to whether the bid should be broken down into parts because of a budgetary concern or be one lump-sum bid.

Scope

For reports about a project in which the design is not included, scope may be an appropriate section in conjunction with the summary instead of an elements of design section. The summary could be everything described in the preceding paragraphs except the listing of the work that was performed during the project, the scope of services. This scope can then translate to areas of focus as described in the following paragraphs. This section would look very much like the scope of service section of the proposal for the project.

Methodology

If the scope section just described is used in the report, methodology is the next logical section that would be included. Methodology is the detailed step-by-step process that took place during the project, from what was read and reviewed, who was interviewed, what was observed, etc. This section would look very much like the methodology section of the proposal for the project.

Security Inventory

The security inventory section is certainly included in a lot of management security reports but also can be in a physical security report as well. It is a listing of all the existing security countermeasures within the facility. This listing can include personnel, physical, and policies and procedures security countermeasures. For a physical security report, the listing and description of the existing physical security countermeasures should be very detailed because that will add to and enhance the recommendations in the elements of design and product selection sections of the report. Obviously, if the project is to review the effectiveness and functionality of existing physical security countermeasure systems, this section will be the majority of the report and would include complementary sections with test reports and functionality results.

Assets

Listing the assets that the designed physical security countermeasures will be protecting may also be appropriate. This section probably would be included only if part of the scope of services for the project was to provide a formal asset analysis or if something in particular that was unexpected by the consultant or client was uncovered during the site visit and observations that needed to be addressed with a security countermeasure.

Threats

A threat assessment or threats section is usually done in a security management report. However, it may also be appropriate for a physical security report if the threat is specific in nature and therefore the physical security countermeasures recommended are to combat that specific threat. Also, if the existence of the specific threats was part of the baseline for the project scope, they may be listed and explained here or in the executive summary.

Vulnerable Areas

The vulnerable areas section should be included in most reports, although the information may be incorporated into another section such as observations. This section describes the vulnerable areas within the facility to protect against the threats and, therefore, why the physical security countermeasures are designed as they are to combat those vulnerabilities. These vulnerable areas should be described in detail with an explanation as to why they are vulnerable and have some connection to the respective countermeasure.

Interview Results

Including the results of the assessment interviews is a powerful way to get a message across to the client. The project then becomes not only what the consultant is saying as the outside expert, but what people within the client's organization are saying as well. The interview results section helps verify the described threats and vulnerabilities and adds credibility to the consultant's recommendations. It gives a sense of ownership to the client for the results and recommendations within the report because the client feels the staff participated in the process, which they did. This is why the consultant should take good notes or tape record the interviews and keep straight who said what. Many times, it's enough just to list the names and titles of those interviewed and a generic list of issues discussed with all the interviewees. Sometimes the report can include anecdotal interview results and comments without authorship, and sometimes the comments can or should be attributed to the specific person.

Observations

In the observations section, the consultant describes what was observed during the site visit survey and observations. These observations should also be very detailed. For instance, the consultant should not say, "I saw a person coming into the facility after hours" but instead say, "At

10:05 p.m. on October 24, I saw a man enter the facility through the bay door at the rear loading dock and go into the shipping area." This detail gives credibility to the assessment and therefore credibility to the report recommendations. It makes the client understand that the consultant was not just walking or standing around doing nothing, but was really observing the facility.

Approvals/Decisions

For physical security countermeasure system projects, the client probably will have some approvals or decisions to make after the assessment report and before the design specifications are written. These decisions will mostly include design, product choice, and division of construction responsibilities issues. These approvals and/or decisions need to be listed in the assessment report in a clear manner so the client knows exactly what needs to be done before the next phase of the project begins. It is also a good idea to include a time frame in which those approvals and/or decisions need to be made in order for the project to stay on the agreed-upon timeline.

Analysis and Recommendations/Conclusions

The analysis and recommendations or conclusions section describes the end result of the report based on the information written in the other sections. It can stand on its own or be incorporated into other sections of the report. If on its own, it should provide a logical list of numbered recommendations with an analysis before or after each recommendation explaining what brought about that conclusion or recommendation. Recommendations or conclusions need to be definitive, not loose or open ended. The consultant is the expert and is supposed to know what to do, so he should say so in no uncertain terms.

Areas of Focus

The report can be broken down into sections called *areas of focus*, which are, in fact, the scope of services. Each area of focus can then include parts of the interview results, observations, vulnerable areas, approvals/decisions, and analysis and recommendations/conclusions sections that pertain to the area of focus. Using an area of focus section is just a different way to format the same information that may (or may not) be clearer for the reader.

Task List

In addition to the client possibly having to make approvals or decisions, there also may be tasks or action items that the client or consultant needs to complete as the next step of the project. As with the decisions section, they should be listed in a clear and logical manner in the task list section, and possibly have a time of completion associated with them to keep the project on schedule.

Attachments

Physical security assessment reports commonly have attachments with them. It would be too cumbersome to have some of this information within the body of the report. Additionally, not everyone reading the report will care about the information in the attachments. Attachments can include the following:

- Door schedule
- Equipment lists
- Conceptual drawings
- Pictures
- Specification sheets
- Software
- Existing equipment functionality reports or test reports

Any attachment that is added to the report should bring added value to the report and to the project. Attachments that don't have a specific purpose should not be added to the report just to make it look good; they will just diminish the rest of the report and the recommendations. The consultant should make sure all the attachments are listed in the executive summary or at the end of the report.

Protection Language

For some clients or projects, including some protection language for the client and consultant within the report may be appropriate. For instance, the report might say the following:

> "It is understood that not all of the recommendations in this report will be adopted for various reasons, including available funding. The failure of *XXXX* to implement recommendations contained in this report should not necessarily be interpreted as a dereliction of duty on the part of *XXXXX*."

The consultant report will more than likely become a permanent record of the company, and the consultant doesn't want to have something in the report come back to haunt the client. This type of language certainly doesn't have to be put in every report, but it is appropriate for some.

DESIGN SPECIFICATIONS

Design specifications and their associated drawings are the documents that the contractor uses as the instructions to perform the installation. They are the expression of the system design in a set format used within the construction and security industries. They are, in fact, the only tool the contractor has to reference how the system should be installed. Whether or not the specifications are accurate and detailed in large part determines whether the installation will go well. That is not to say a good contractor cannot complete a good installation without good specifications, but even if the contractor does, the system may not be installed exactly the way it was supposed to be if the design specifications are poor.

It cannot be stressed enough that design specifications must be of great QUALITY. The absolute garbage seen in the field, mostly produced by installation companies and architects, passed off as design specifications is atrocious and the reason the industry needs more professional independent security design consultants. The section on quality in Chapter 1 listed some common garbage examples, such as specifications and drawings not matching and a different company's name actually being listed within the specifications. A professional security design consultant's work is simply superior to this. Appendix D shows actual design specifications that were used for an IFB in conjunction with the associated drawings in Appendix E. One of the more rewarding professional moments in this author's career was when a well-respected integrator called these exact specifications the "best specifications he had ever seen." Praise like that will come more than once with the continuation of excellent specification writing.

Every security design consultant has his own style of language he uses in his design specifications, each conveying the same message using different words. While it is commonplace and certainly not unethical to take some specific words or concepts from other specifications and incorporate it into the consultant's language, it is unethical to directly copy large amounts of language from someone else. There probably are some legal complications there as well. The more specifications the consultant reads, the more he gets a feel for what is good or bad about the writing and the more refined the language is in his own specifications.

Design specifications are separated into three main parts: General, Products, and Execution. These are industry standards and should never be altered. Each part is broken down into many sections, which can be altered slightly based on the project. If the consultant is working for an architect or the project is part of a larger construction project, there may also be specific part and section numbers that must be associated with the respective sections. These are industry standard section numbers used in the architectural and construction fields. They will not be discussed in detail here and are not included within the specifications in Appendix D. While specifications can certainly include more sections than are described here based on the complexity of the project, they should almost never have less than what is described. The sections may also have different names than described here. Following is a description of the parts and sections.

Part I: General

Part I of the specifications includes all the provisions of the specifications that are not product or installation technique oriented. This part includes all the *administrative provisions* for lack of a better term, although they are just as important as the rest of the specifications. They include the items described in the following sections.

Description The description section, also called Project Summary, provides the overall description of the project and synopsis of the specifications, similar to an executive summary of a report. It should include the project address and other pertinent information, an overview of the equipment that will be installed, and a general breakdown of the contractor's responsibilities regarding the project. It can also include the client's responsibilities regarding the project.

Definitions Terms that are used throughout the specifications are defined in the definitions section. The most common definitions include *Contractor, Consultant, Owner*, and *Work*. There are good specifications that include many more definitions than this, but in most cases that is not necessary. Throughout the specifications, if the first letters of the particular words listed in the definitions section are capitalized, it means those words are as defined in this section.

Overview The overview, also called "work included," is a more detailed description of the scope of the work than is included within the description section. It may also include some items of work not included at all in the description section.

Related Documents The related documents section provides a reference to other documents that are part of or relate to these specifications. The design drawings are the most common example. Drawings should be listed with their drawing number and drawing name.

Intent The intent section is there to cover all the bases in case of minor discrepancies or errors within the specifications or drawings. It is basically a set of statements about the intent of the project work and the contractor's responsibilities.

Contractor Qualifications The contractor qualifications section is extremely important because it outlines the necessary qualifications of the contractor to even bid on the project as well as perform the work. The qualifications need to be very specific and stringent to the degree appropriate for the project. Qualifications can include licensing, training, experience, good references, proof of insurance, length of time in business, litigation record, etc. It should also be noted in this section that any bidders must provide this information or be considered unresponsive and their bid unacceptable.

Submittals The submittal requirements of the contractor before the installation begins and before it is deemed complete are described in the submittals section. The submittals include shop drawings, technical data sheets, functional diagrams, and as-built drawings. These items are important because they describe exactly how the contractor will or has installed the system, and it's the consultant's job to make sure the installation is correct. The consultant's requirements of the contractor for submittals should be very detailed for that reason. The consultant should always ask the contractor to use the same symbol set for the shop drawings and as-built drawings as are used on the design drawings for consistency for all parties, especially the client. This section, or another one called proposal, may also include the details of what is required in the submitted proposal from the contractor.

Arrangement of Work The arrangement of work section is another "cover all the bases" section dealing with device locations. This type of section may seem silly, until a contractor installs a device in the wrong location and blames the consultant.

Project Conferences The specifications need to spell out the exact responsibilities of the contractor with regard to project conferences. The number and types of conferences should be listed and described along with who will be present. Contractors generally hate these conferences, so the specifications should make it clear that they are mandatory along with all the other details.

Subletting If *subletting,* also called *subcontracting,* is allowed for the project, the subletting section describes the allowed type and amount of subletting, the qualifications of the subcontractors, and their relationship to the contractor and project. The contractor should always be responsible for the subcontractors' work. If no subletting is allowed, this section will state that.

Verifying Jobsite Conditions The jobsite conditions section tells the bidders that they are responsible for looking at the facility themselves before submitting bids, and if they are awarded the contract, ignorance of the facility is not an excuse for any work problems or incorrect bidding.

Changes in Scope The changes in scope section describes how any changes in the scope of the project will be communicated and dealt with during the course of the installation.

Product Handlings The product handlings section describes how the products and materials for the project will be stored at the site, who will provide this storage, and who is responsible for the security of those materials and products. In addition, this section details what happens if any of the stored materials or products are damaged.

Security The security section details what happens and who is responsible if any of the stored materials or products are stolen or lost.

Temporary Security/Protection The safety precautions within the facility that the contractor is responsible for during the installation are described in the temporary security/protection section. Such things as hard hats, cones, barriers, etc., are described in detail. Any client responsibilities for safety are also described.

Guaranty The guaranty, or warranty, section defines the guaranty the contractor will provide for the installed system, including the length of time of that guaranty, when that time frame begins, and what the guaranty will cover. The language of this section needs to be written carefully so there are no loopholes.

Defects The defects section describes the acceptable condition of the material and products furnished by the contractor and the process that will take place if any of the items are deemed unacceptable, along with who will make that decision.

Reference Standards All the applicable codes, ordinances, standards, and guidelines that the contractor must follow for the installation are listed in the reference standards section.

Equipment Substitutions If equipment substitutions are allowed for the project, the equipment substitutions section details the exact criteria that must be met for the proposed substitute product to be accepted. It will also describe the time frame the bidders have to submit such proposed substitutions and who will make the decision about that acceptance or nonacceptance, along with how the determination will be made. If no substitutions are allowed for the project, this section will state that.

Owner Training The contractor needs to provide defined training to the owner at the end of the installation. The number of hours of training, who will provide it, what it will include, and when it will take place all need to be included in the owner training section.

Part II: Products

The products part of the specifications includes all the required information about the products for the system or systems. The sections include the items described next.

System Components The system components section is just an overview of the expected quality and functionality of the specified products.

System Specifications By far the longest section of the specifications, the system specifications section lists all the products that will be used in the system (if an IFB) and includes all the Architects and Engineers (A&E) Specifications for those products (for RFP or IFB). A&E Specifications are very detailed point-by-point descriptions of the physical and functional elements of the products. Many manufacturers already have A&E Specifications written for their products. If this is the case, the consultant needs to reformat them to be consistent with the rest of the specifications and add the specific language to them appropriate for the project or change some of the language as necessary. This long and tedious process is not much fun. Many of these specifications are very lengthy and dull to read and write. However, once this section is completed for a product, if that product is used again on another project, only minor changes to the original one would need to be made.

If the manufacturer does not have A&E Specifications written for its products, which is common, then the consultant must write them. The easiest way to do this is to take the technical data sheet for the product and go point by point down the data sheet and write the specification. Again, this is a long and tedious process. These A&E Specifications should be grouped by system and include every product that makes up the system. If any product is left out, the decision on what will be used now will fall in most cases to the contractor, which is not where it should be. The consultant is being paid and has the expertise to make all those decisions. He should not leave anything out just because it's a relatively small product within the system.

Part III: Execution

The execution part of the specifications includes all the detailed information about the installation itself in terms of function and quality. The overall quality of the installation is defined in many ways by the quality of the installation descriptions in this part. The sections of this part include the items described next.

Engineering and Design The engineering and design section is another "cover all the bases" section that in short says the contractor is responsible for making the system function as it is intended. It also tells the contractor how to rectify discrepancies within the specifications or drawings. A professional consultant strives very hard not to have any of these discrepancies, but mistakes can always be made.

Testing and Inspection The testing and inspection section, also called commissioning, describes the testing process at the end of the installation, including how the test will be conducted, when it will be conducted, who will be present for the test, and how the test reports and as-built drawings will be produced and reviewed.

Conduit and Cabling The conduit and cabling section details the types of cabling and conduits that will be used for the project and how they will be installed. These are very specific details expressing the quality that is expected. It is not enough to say, "Follow the code." If the consultant wants the wiring or conduit installed in a certain manner, that should be outlined in this section. This section cannot have too much detail.

Operational Objectives The specifications must explain to the contractor what the end result of the system needs to be, exactly how is it supposed to work. Although including this information may seem obvious, it is left out of many less quality specifications. The operational objectives section should provide a step-by-step description of how each system should operate based on certain conditions. Again, this section should be very detailed, taking into account every action that can be anticipated in the function of the system.

Scope of Work: Specific Installation Criteria The scope of work section provides a listing of all the products grouped by system and the specific installation criteria the consultant wants the contractor to utilize for the installation. It includes exact mounting locations with dimensions, if appropriate, mounting or installation methods, and electrical requirements. Again, specificity is the key.

Cutting and Patching The cutting and patching section gives the installation requirements for cutting into or penetrating the structure of the facility and patching those penetrations. This is for flooring, walls, or any other pertinent structure.

Manuals and Software The manuals and software section provides a description of all the manuals, software, drawings, etc., that the contractor will supply to the owner at the end of the project. It describes the number of sets required, how they will be packaged together and collated, and when they will be delivered.

Drawings

Drawings are the complement to the written design specifications and are a diagrammatic expression of the design. They are used both to help the bidders prepare bids and for the contractor on site to perform the installation. As with the written specifications, there are a lot of garbage drawings in the industry. The most important issue of QUALITY for drawings is that the drawings must match the specifications. In other words, the written specifications and drawings must say the same thing; there can be no contradictions. Nothing is more frustrating for an integrator or client than the specifications saying two different things in two areas.

Appendix E shows actual design drawings that were used in conjunction with the written specifications in Appendix D. While the layout and format of the drawings are a matter of both personal taste and the skill of the AutoCAD operator, certain types of drawings are required for certain types of physical security countermeasure system projects. They can include the items described in the following sections but may also include more not described here.

Cover Sheet The cover sheet is the introductory page giving the client name, name of project, consultant's company name, and an index of all the sheets or drawings in the design package. The cover sheet may also include a legend of the symbols used on all the drawings specific to the designed systems. This legend may also be included on a different sheet. Regarding the symbol legend, only the symbols that are used on the drawings should be included in the legend. It is common practice to include all symbols used by the consultant on the legend, regardless of whether they are used on the drawings, because then the legend has to be made only once. However, showing all these symbols can become confusing to everyone reading the drawings and, in the opinion of this author, is lazy and unprofessional.

Numerous symbols are used in the industry to identify the exact same piece of equipment. It is not uncommon to find two or three different symbols used for the same equipment by different people on the same project. Over the years there has not been a standard set of security symbols used in the industry, unlike fire system symbols. This is changing with the development of an Architectural Graphics Standard–CAD Symbols for Security System Layout by the SIA/IAPSC, a collaborative effort between the Security Industry Association (SIA) and the International Association of Professional Security Consultants (IAPSC). These symbols in written and electronic form can be found at siaonline.org. These symbols should be used on all physical security countermeasure system drawings so that over time there will be a standard set used in the industry.

Floor Plan The floor plan is the layout of the facility with the location of all the devices used for the systems represented by symbols on the drawing. The floor plan should show all the rooms of the facility, so it is a true representation of the layout of the facility and the facility can be navigated using the drawing. All field devices and head-end equipment should be expressed on the drawing. While in many cases putting the symbol in the exact location of the device is impossible, it should be as close as possible. Using arrows or some kind of insertion point symbol is a help. All the drawings after the cover sheet have a title block of some kind with the sheet name and other information from the cover sheet.

Door Schedule A door schedule may be appropriate, especially for drawings of the design of an access control system. Door details or door elevations may also be included on this sheet or a separate detail sheet.

Riser Diagrams Consultants have many interpretations of what a riser diagram should look like, but the common factor is that all versions show the physical relationship between the equipment of the system. Such a diagram may or may not show the type of wiring between the devices and head-end equipment. They are very important, so every set of drawings should include a riser diagram for each system. While the riser diagram may not show the exact physical location of each device, it does show which floor the device is located on for a multifloor facility.

Details There are many kinds of detail sheets possible based on the type of system, including panel layout detail, door detail, specific device detail, equipment elevation detail, console elevation detail, etc. The details drawn should be those that are necessary for the contractor to properly install the system.

11 Project Management

The third set of services provided by a security design consultant is project management services. Sometimes these services are called *construction administration* or *contract administration* as well. These services take place during the procurement and construction phases of a physical security countermeasure installation project. Basically, project management is making sure a qualified contractor installs the system as it was designed and that it functions correctly after it is installed.

The project management services that will be performed by a security design consultant on a project vary from project to project and client to client. The scope of the project management services to be provided needs to be clearly defined with the client before a proposal for services can be submitted. Some projects warrant only some of the available project management services, whereas others warrant almost all of them. Many clients, especially municipal and government clients, will also have an in-house project manager for the project to oversee the installation and be the contact person for the client. Even if there is no official project manager, there is always someone representing the client overseeing the installation. This does not mean that no project management services are required from the consultant—quite the contrary. The client's project manager probably has no expertise on physical security countermeasures, including how they should be installed. This person also doesn't have as firm a grasp of the specifications as the consultant. Conversely, the consultant probably cannot be on site all the time and cannot independently coordinate the necessary assets at the facility needed for a successful installation. Therefore, the two have complementary functions. The interaction and communication between the consultant and client's project manager are key to a successful installation. The following sections describe the project management services that can be provided.

ASSIST IN ATTACHING BID DOCUMENTS TO THE DESIGN SPECIFICATIONS

The written design specifications and drawings cannot be put out to bid by themselves. Bid documents need to be attached giving all the bidding instructions, legal and contract issues, bid forms, etc. For virtually all municipal and government clients and most private clients, these documents already exist and just need to be updated with the specifics of the project. Sometimes clients will simply take the consultant's specifications and add them to the bid package themselves. Others want the consultant to alter the bid documents for the project and prepare a complete package for client review. This task can be time consuming if the consultant isn't familiar with the bid documents, so he must make sure the details of this service are clearly defined within his proposal.

If the documents are being put together by the client, the consultant needs to make sure to review all the bid documents before releasing to prospective bidders. Sometimes language within the client's documents will contradict something in the design specifications. Contractor qualifications, submittals, and warranty are common discrepancies. These discrepancies need to be worked out and fixed before they are released. If the client has no bid documents and wants the consultant to write them, he should make them as simple and concise as possible. Instructions to bidders and a bid form may be enough to add to the design specifications. The consultant should make sure the client reviews and approves what is written before the bid is released.

ASSIST IN LOCATING AND INVITING QUALIFIED CONTRACTORS TO BID

Many times the client has no idea who should bid on the installation of the physical security countermeasure system. It is in the best interest of the client and therefore the consultant that qualified contractors bid on the project. The consultant should prequalify two or three contractors that will submit a bid. The consultant can find them to perform this prequalification in a few ways:

- Contact the manufacturer(s) of the specified equipment and ask if there are "preferred" or "trained" contractors (sometimes called *business partners*) in the appropriate geographical area and then contact them to see if they are interested in bidding on the project. Make sure the specifics of the system are described so they give an accurate answer about bidding.
- Contact distributors in that geographic area to see if they know of qualified contractors and then contact them as indicated in the preceding method. A short conversation with a company representative will make it clear whether or not the company knows what it's doing and is capable of installing the specified system. This is another reason to develop relationships with distributors.
- Check with other contractors that the consultant has had good experiences with and ask if they know someone in that area. They just might.
- If all else fails, use an Internet search engine to come up with a list of potential contractors and then make phone calls.

Even if the project is an open bid and will "hit the street," the consultant should still give a couple of contractor names to the client so an invitation to bid on the project can be sent. These contractors may not see the bid otherwise, and there is a risk of not having the best qualified contractors bidding.

LEAD THE PRE-BID CONFERENCE AND WALK-THROUGH

Many, if not most, projects start with a pre-bid conference and walk-through. At this point, the project bidders, client, and consultant get together so the contractors can see the facility and ask any questions. In theory, the contractors should have acquired and reviewed the specifications before attending the conference, but in reality that is not always the case. Just as it's very difficult in some cases for a consultant to bid on a project without seeing the facility first, the same goes for contractors, probably even more so. These conferences usually start in a room where the contractors can ask any questions of the consultant or client, the client can go over any legal issues, etc., and the consultant can briefly review the specifications. Then the facility is walked through so the contractors can see what is described in the specifications and familiarize themselves with the size and construction of the facility. The consultant should lead this walk-through and actually point out all the pertinent device and head-end locations and any notable construction issues.

There is some difference of opinion within the security design consulting field as to whether these pre-bid conferences are worth the effort. This author believes they are. The key is to make them short and informative. The purpose should be about the contractors getting all the information they need and questions answered so they can submit the best bids possible. Contractors will always bid high if they don't have all the facts. The conference is not about the consultant or client regurgitating the specifications orally or reciting information everyone already knows.

Some clients can make the pre-bid conference and walk-through mandatory for contractors to be qualified to submit bids. Some municipal and government clients cannot do that by law. Instead, they say the walk-through is "strongly recommended." Unless the project is very small or there is no facility yet to visit, the conferences should be mandatory whenever possible.

PROVIDE WRITTEN ANSWERS TO QUESTIONS AT PRE-BID WALK-THROUGH

During the pre-bid conference and walk-through, the contractors will have questions. Most of those questions can be answered right there at that time. The consultant should make sure everyone in attendance hears the questions and responses. If the conference is not mandatory, it is usually not necessary to inform the nonattending bid package holders of those questions and answers. If the consultant cannot answer some questions at that time, he needs to put those questions and answers in writing after the fact and distribute them to every conference attendee. In some cases other bid package holders not present also need that information.

The consultant should not be afraid to answer a question with "I don't know." It is far better to say that and give the proper answer later than to be wrong with the answer at the time. In any event, the consultant should make sure to know the specifications well before the conference so the wrong information isn't given even accidentally. There may have been substantial time between when the specifications were written and the pre-bid conference occurred, so a quick review before the conference is a good idea.

APPROVE PRODUCT CHANGES

Even for an Invitation For Bid (IFB), the client may either be forced or want to allow for equipment substitutions or "approved equals," as long as they meet the exact same criteria as the specified equipment. If this is the case, the bidders usually have a set amount of time that begins after the pre-bid conference and ends before the bids are due in order to submit requested substitutions for approval. The consultant needs to review any of those requests and submit to the client in writing whether a substitution is acceptable and why. If the substitution is acceptable, then the client needs to inform all the bidders of that fact so everyone is on the same level playing field. If the specifications are written correctly, there should be no need for such substitutions, and they should be at least tacitly discouraged.

ASSIST IN EVALUATION OF BIDS

Once the client receives the submitted bids, they need to be reviewed by both the client and the consultant. In some cases the consultant actually receives the bids and then ranks them for the client, but that is rare. The first item the consultant should review is contractor qualifications. Very specific contractor qualifications should have been written into the specifications, and the consultant needs to make sure the contractors meet those qualifications. This includes checking references by making phone calls, checking licenses, etc. If the contractors do not meet the qualifications, they should not be considered for the awarded contract in most cases. If some clients have flexibility in awarding contracts (mostly private only) and only minor information is missing, the consultant may contact the bidder for that information. The consultant needs to make sure there is the authority from the client to contact any bidder before the contract is awarded so no laws or client policies are broken.

If the bid is an IFB, the consultant should review the rest of the bid package to make sure everything is in order before a recommendation is made. The client will likely do most of the noncontractor qualification reviewing, but two sets of eyes are better. If the bid is a Request For Proposal (RFP), then the consultant has to perform an evaluation of the whole bid package regarding product choices, how they will be integrated, etc., to make sure the system and products chosen by the contractor will meet the needs of the client and satisfy the criteria of the specifications in every way. This is obviously a more comprehensive review that takes longer and in some cases is done in conjunction with the client in a formal setting. In either case, the consultant needs to make a formal recommendation in writing (or e-mail) with what contractor is recommended to be awarded the contract or a short-ranked list of two or three contractors.

REVIEW AND APPROVE ALL SUBMITTALS

For most projects, the contractor will be required to provide submittals to the consultant for approval after being awarded the contract but before any work begins. These submittals include shop drawings, wiring diagrams, and product technical data sheets. The purpose of these submittals is to verify to the client and consultant that the contractor understands how to install the designed system and to provide a baseline starting point of the installation for the contractor.

Appendix F shows an actual set of as-built drawings from a contractor for a project. The shop drawings are the first stage of as-built drawings and contain as much information as is known at the start of the project. Shop drawings MUST include wiring point-to-point diagrams, a riser diagram, and a cable legend. The idea is for the consultant to verify the contractor will wire the system correctly with the correct wire size and type. Shop drawings are usually submitted using AutoCAD, although sometimes contractors will convert the drawings from AutoCAD to PDF (Adobe Acrobat) files for proprietary reasons.

Technical data sheets are provided by manufacturers describing the specifications of a particular piece of equipment. Data sheets for all equipment that will be part of the installation must be provided to ensure the correct equipment will be utilized. Additional wiring diagrams may include block diagrams with the interaction of system components, system functional diagrams, or other such necessary information depending on the project.

The consultant should review these submittals very carefully and pay attention to every detail to make sure everything is correct. If it is not correct, the consultant and contractor can communicate and fix whatever is wrong. Since this is a detailed description of how the installation will take place, the project will go much smoother if the submittals are correct before work begins. This is one of the more critical project management functions for the consultant. Once the submittals have been approved, the consultant can inform the contractor and client that the job is ready to proceed. Because the project cannot proceed without consultant approval of the submittals, the consultant needs to review and approve or disapprove the submittals in a timely manner.

RESPOND TO ANY CONTRACTORS' REQUESTS FOR INFORMATION

During the construction process, if the contractor has any technical or installation questions about the design specifications or drawings, the consultant needs to answer those questions. Sometimes this is a formal process in writing with a specific form, but most of the time it is just a phone call or e-mail. How this information is disseminated depends on the project type, client, and contractor. The consultant should make sure these questions and answers are documented and kept in some way, so if there are any future problems about an already addressed issue, the correspondence can be referenced.

It is important for the consultant to develop a good relationship with the contractor so the contractor feels comfortable asking questions to make sure everyone is on the same page. The idea is to create a partnership between the consultant, contractor, and client to successfully complete the installation on time and within budget. However, the consultant represents the client when dealing with the contractor, and he may have to be firm at times, so the consultant must maintain a professional relationship with the contractor at all times.

ATTEND CONSTRUCTION MEETINGS

During the course of the installation, there will be construction meetings to review the details of the project, schedule updates, and address coordination issues. It is necessary for the consultant to be at some or all of those meetings. The first one is usually a construction kick-off meeting where all the parties get together, the contractor gets a "notice to proceed," and scheduling and other issues are discussed. The consultant should definitely attend this meeting to answer any questions and make sure everything is proceeding according to the schedule and specifications. The consultant and client's project manager can discuss and agree on their respective roles and expectations at this meeting. This is also the opportunity for the consultant to inform the contractor of what is expected regarding the quality and details of the installation. There may be other formal meetings throughout the construction as well, depending on the size and scope of the project. Some larger construction projects meet weekly. The consultant does not necessarily have to attend all these meetings; his attendance depends on the geographic location of the project, what was proposed for services, type of installation, etc. What meetings will be attended should be agreed upon in advance with the client's project manager.

CHECK, REVIEW, AND APPROVE INSTALLATION PROGRESS AND TECHNIQUES

The consultant's job is to make sure the designed physical security countermeasure system is installed according to the specifications. In addition, the consultant needs to make sure proper installation techniques are used and all appropriate codes are followed. The only way to accomplish this is to periodically check the job during the installation. It's easy to tell whether devices are installed and, therefore, where the project is regarding the schedule; it is more difficult to determine if the equipment is installed correctly. This is where industry experience is a must for the security design consultant. It's impossible to know if something is installed correctly if the consultant does not know what is and isn't correct.

The consultant should review at least a sampling of each type of installed device. For instance, for an access control system, taking two or three readers off the wall to make sure there are wire tags on the incoming wires and the blocking for the reader is correct would be appropriate. Not every one would necessarily have to be checked. All head-end equipment should be checked. Any deficiencies need to be pointed out to the contractor so they can be corrected. These items then need to be checked again during that site visit or the next one. There is no set rule on the number of visits by the consultant to check the project installation progress and quality. Again, the number of visits depends on the type and size of the project, geographic location, and scope of services proposed. For most projects, a minimum of two visits is required for a proper review, although making five visits or more is not uncommon.

REVIEW AND APPROVE ANY CHANGE ORDERS OR CHANGE IN SCOPE OF WORK

Although a goal of the security design consultant is to have no change orders during the installation of a designed system, that does not always happen. In many cases it is no fault of the design but of changes to the project as it progresses. In some cases the consultant may make a change because of an unanticipated problem. In either case, if additional or different work or equipment needs to be provided by the contractor, a change order needs to be prepared and submitted for approval. It can be directly submitted to the consultant or the client. The consultant should review and either approve or deny the change order and forward the decision and reasoning to the client. Denial of a change order may be due to scope of work or equipment or cost. Communication between the consultant and contractor should take place before the change order is submitted to ensure everyone is on the same page. Change orders should always be in writing, not just verbal. The client can then add the change of scope and price to the contract with the contractor.

WITNESS SYSTEM TEST WITH CONTRACTOR

Testing the system is a critical element of the installation process. One would think that making sure every element of the system works properly before commissioning the system and turning it over to the client is common sense, yet it is shocking how many systems are installed in the field and operating with no one having any idea if it is all working correctly. The consultant must witness the testing of the system to assure the client that the installed system works in every way according to the specifications. The testing of the system must be a FULL test. In other words, every device and piece of head-end equipment must be tested for each condition that will be presented in real operations. Some designers or project managers perform a test of half or even fewer of the field devices and assume the rest must be working correctly. That is nonsense and certainly not professional or industry best practices. If anything does not test correctly, it must be fixed by the contractor and retested. The consultant cannot assume that it will test correctly after being fixed. Whether the consultant needs to be present for the subsequent tests depends on the individual project, status of the client's project manager, and history with the contractor. All testing should be coordinated between contractor and consultant well in advance of the testing date to make sure enough time is set aside and proper personnel are available for the test.

REVIEW AND APPROVE TEST REPORT

After the testing is complete, the contractor needs to submit a test report outlining all the devices and head-end equipment that were tested and the test results. This test report is used as proof to the client that the system is working according to the specifications. If that turns out not to be the case, the client has recourse based on the test report. It also forces the contractor, along with the consultant being present during the testing, to make sure everything is working properly. The consultant needs to review and approve this test report before signing off that the installation is complete.

Tables 11.1 and 11.2 provide examples of test reports. The first is a fairly comprehensive report that also includes wire type and other information not directly related to the test. The second is a less comprehensive

Table 11.1 Comprehensive Test Report

ABC Company **1234 Main St.** **Anywhere, USA 12345**

AB—Technician #1

CD—Technician #2

INPUTS		ACU 1								
LOC.		DESCRIPTION	I/O TYPE	WIRE TYPE	WIRE NO.	POINT CONFIGURATION	TERMI-NATED PANEL	FIELD	SOFT-WARE	TURN-OCDR
0	**ACU1**	East Glassbreaks	SI	22/4	GB1	N/O Series Parallel	AB	CD	CD	AB & CD
1	**ACU1**	West Glassbreaks	SI	22/4	GB2	N/O Series Parallel	AB	CD	CD	AB & CD
2	**ACU1**	Electrical Rm Temp Sensor 1	SI	22/4	TS1	N/C Series Parallel	AB	CD	CD	AB & CD
3	**ACU1**	Equipment Rm Temp Sensor 2	SI	22/4	TS2	N/C Series Parallel	AB	CD	CD	AB & CD
4	**ACU1**	Remote Release	SI	22/4	DR	N/O Series Parallel	AB	AB	AB	AB & CD
5	**ACU1**	SPARE	SI	22/4						
6	**ACU1**	SPARE	SI	22/4						
7	**ACU1**	SPARE	SI	22/4						
8	**ACU1**	SPARE	SI	22/4						
9	**ACU1**	SPARE	SI	22/4						
10	**ACU1**	SPARE	SI	22/4						
11	**ACU1**	SPARE	SI	22/4						

RRE4 INPUTS										
16	**RRE4-1-RDR0**	CR-001 Door Contact	SI	22/4	CR001 DC	N/C Series Parallel	AB	AB	CD	AB & CD
17	**RRE4-1-RDR0**	CR-001 REX Motion	SI	22/4	CR001 REX	N/O Series Parallel	AB	AB	CD	AB & CD
18	**RRE4-1-RDR0**	SPARE	SI	22/4						
19	**RRE4-1-RDR0**	SPARE	SI	22/4						
20	**RRE4-1-RDR1**	CR-003 Door Contact	SI	22/4	CR003 DC	N/C Series Parallel	AB	AB	CD	AB & CD
21	**RRE4-1-RDR1**	SPARE	SI	22/4						
22	**RRE4-1-RDR1**	SPARE	SI	22/4						
23	**RRE4-1-RDR1**	SPARE	SI	22/4						
24	**RRE4-1-RDR2**	CR-004 Door Contact	SI	22/4	CR004 DC	N/C Series Parallel	AB	AB	CD	AB & CD
25	**RRE4-1-RDR2**	CR-004 REX Motion	SI	22/4	CR004 REX	N/O Series Parallel	AB	AB	CD	AB & CD
26	**RRE4-1-RDR2**	SPARE	SI	22/4						

Continued

Table 11.1 *Continued*

27	**RRE4-1-RDR2**	SPARE	SI	22/4						
28	**RRE4-1-RDR3**	CR-005 Door Contact	SI	22/4	CR005 DC	N/C Series Parallel	AB	AB	CD	AB & CD
29	**RRE4-1-RDR3**	CR-005 REX Motion	SI	22/4	CR005 REX	N/O Series Parallel	AB	AB	CD	AB & CD
30	**RRE4-1-RDR3**	SPARE	SI	22/4						
31	**RRE4-1-RDR3**	SPARE	SI	22/4						
32	**RRE4-1-RDR0**	CR-006 Door Contact	SI	22/4	CR006 DC	N/C Series Parallel	AB	CD	CD	AB & CD
33	**RRE4-2-RDR0**	SPARE	SI	22/4						
34	**RRE4-2-RDR0**	SPARE	SI	22/4						
35	**RRE4-2-RDR0**	SPARE	SI	22/4						
36	**RRE4-2-RDR1**	CR-013 Door Contact	SI	22/4	CR013 DC	N/C Series Parallel	AB	CD	CD	AB & CD

37	**RRE4-2-RDR1**	SPARE	SI	22/4						
38	**RRE4-2-RDR1**	SPARE	SI	22/4						
39	**RRE4-2-RDR1**	SPARE	SI	22/4						
40	**RRE4-2-RDR2**	CR-014 Door Contact	SI	22/4	CR014 DC	N/C Series Parallel	AB	CD	CD	AB & CD
41	**RRE4-2-RDR2**	CR-014 REX Motion	SI	22/4	CR014 REX	N/O Series Parallel	AB	CD	CD	AB & CD
42	**RRE4-2-RDR2**	SPARE	SI	22/4						
43	**RRE4-2-RDR2**	SPARE	SI	22/4						
44	**RRE4-2-RDR3**	CR-017 Door Contact	SI	22/4	CR017 DC	N/C Series Parallel	AB	CD	CD	AB & CD
45	**RRE4-2-RDR3**	SPARE	SI	22/4						
46	**RRE4-2-RDR3**	SPARE	SI	22/4						
47	**RRE4-2-RDR3**	SPARE	SI	22/4						

Continued

Table 11.1 *Continued*

READERS										
0	**RRE4-1-RDR0**	Main Entry CR-001 Reader	RDR	18/6 SH	CR001 CR	Weigand	AB	AB	CD	AB & CD
1	**RRE4-1-RDR1**	Lobby CR-003 Reader	RDR	18/6 SH	CR003 CR	Weigand	AB	AB	CD	AB & CD
2	**RRE4-1-RDR2**	Employee 1 Entry CR-004 Reader	RDR	18/6 SH	CR004 CR	Weigand	AB	AB	CD	AB & CD
3	**RRE4-1-RDR3**	Courier Entry CR-005 Reader	RDR	18/6 SH	CR005 CR	Weigand	AB	AB	CD	AB & CD
4	**RRE4-2-RDR0**	Print/Pack CR-006 Reader	RDR	18/6 SH	CR006 CR	Weigand	AB	CD	CD	AB & CD
5	**RRE4-2-RDR1**	Electrical Rm CR-013 Reader	RDR	18/6 SH	CR013 CR	Weigand	AB	CD	CD	AB & CD
6	**RRE4-2-RDR2**	Employee 2 Entry CR-014 Reader	RDR	18/6 SH	CR014 CR	Weigand	AB	CD	CD	AB & CD
7	**RRE4-2-RDR3**	Equipment Rm CR-017 Reader	RDR	18/6 SH	CR017 CR	Weigand	AB	CD	CD	AB & CD

OUTPUTS										
0	**ACU1**	Vista 50 Zone 1	DO	18/2		N/O Parallel	CD	CD	AB	AB & CD
1	**ACU1**	Vista 50 Zone 2	DO	18/2		N/O Parallel	CD	CD	AB	AB & CD
2	**ACU1**	Vista 50 Zone 3	DO	18/2		N/O Parallel	CD	CD	AB	AB & CD
3	**ACU1**	Vista 50 Zone 4	DO	18/2		N/O Parallel	CD	CD	AB	AB & CD
4	**ACU1**	Vista 50 Zone 5	DO	18/2		N/O Parallel	CD	CD	AB	AB & CD
5	**ACU1**	Vista 50 Zone 6	DO	18/2		N/O Parallel	CD	CD	AB	AB & CD
6	**ACU1**	Vista 50 Zone 7	DO	18/2		N/O Parallel	CD	CD	AB	AB & CD
7	**ACU1**	Vista 50 Zone 8	DO	18/2		N/O Parallel	CD	CD	AB	AB & CD
8	**ACU1**	SPARE	DO	18/2						
9	**ACU1**	SPARE	DO	18/2						
10	**ACU1**	SPARE	DO	18/2						
11	**ACU1**	SPARE	DO	18/2						
16	**RRE4-1-RDR0**	CR-001 Electric Handset	DO	18/2	CR001 EH	N/O 24VDC	AB	AB	CD	AB & CD
17	**RRE4-1-RDR0**	SPARE	DO	18/2						
18	**RRE4-1-RDR1**	CR-003 Door Strike	DO	18/2	CR003 ES	N/O 24VDC	AB	AB	CD	AB & CD
19	**RRE4-1-RDR1**	SPARE	DO	18/2						
20	**RRE4-1-RDR2**	CR-004 Door Strike	DO	18/2	CR004 ES	N/O 24VDC	AB	AB	CD	AB & CD
21	**RRE4-1-RDR2**	SPARE	DO	18/2						

Continued

Table 11.1 *Continued*

22	**RRE4-1-RDR3**	CR-005 Door Strike	DO	18/2	CR005 ES	N/O 24VDC	AB	AB	CD	AB & CD
23	**RRE4-1-RDR3**	SPARE	DO	18/2						
24	**RRE4-2-RDR0**	CR-006 Door Strike	DO	18/2	CR006 ES	N/O 24VDC	AB	CD	CD	AB & CD
25	**RRE4-2-RDR0**	SPARE	DO	18/2						
26	**RRE4-2-RDR1**	CR-013 Door Strike	DO	18/2	CR013 ES	N/O 24VDC	AB	CD	CD	AB & CD
27	**RRE4-2-RDR1**	SPARE	DO	18/2						
28	**RRE4-2-RDR2**	CR-014 Door Strike	DO	18/2	CR014 ES	N/O 24VDC	AB	CD	CD	AB & CD
29	**RRE4-2-RDR2**	SPARE	DO	18/2						
30	**RRE4-2-RDR3**	CR-017 Door Strike	DO	18/2	CR017 ES	N/O 24VDC	AB	CD	CD	AB & CD
31	**RRE4-2-RDR3**	SPARE	DO	18/2						

Additional Points									
Hardwired	Receptionist Duress—Vista 50 Zone 10	SI	22/4	PB	N/O Parallel	AB	AB	AB	AB & CD
Hardwired	Receptionist Lock Release		18/2	MR	N/O	AB	AB	AB	AB & CD
Hardwired	Receptionist Keypad		18/2	KP	N/C	AB	AB	AB	AB & CD
Hardwired	Receptionist Electric Strike Dr 011		18/2	ES011	N/O 24VDC	AB	CD	AB	AB & CD
East Glassbreaks Zn 1	Glassbreak #1 interfaced with ACU1 output 04 & Vista 50 Zn 5 (Numbered east to west)					AB	CD	AB	AB & CD
East Glassbreaks Zn 1	Glassbreak #2 interfaced with ACU1 output 04 & Vista 50 Zn 5					AB	CD	AB	AB & CD
East Glassbreaks Zn 1	Glassbreak #3 interfaced with ACU1 output 04 & Vista 50 Zn 5					AB	CD	AB	AB & CD
East Glassbreaks Zn 1	Glassbreak #4 interfaced with ACU1 output 04 & Vista 50 Zn 5					AB	CD	AB	AB & CD
East Glassbreaks Zn 1	Glassbreak #5 interfaced with ACU1 output 04 & Vista 50 Zn 5					AB	CD	AB	AB & CD

Continued

Table 11.1 *Continued*

East Glassbreaks Zn 1	Glassbreak #6 interfaced with ACU1 output 04 & Vista 50 Zn 5	AB	CD	AB	AB & CD
East Glassbreaks Zn 1	Glassbreak #7 interfaced with ACU1 output 04 & Vista 50 Zn 5	AB	AB	AB	AB & CD
West Glassbreaks Zn 2	Glassbreak #8 interfaced with ACU1 output 05 & Vista 50 Zn 6	AB	AB	AB	AB & CD
West Glassbreaks Zn 2	Glassbreak #9 interfaced with ACU1 output 05 & Vista 50 Zn 6	AB	CD	AB	AB & CD
West Glassbreaks Zn 2	Glassbreak #10 interfaced with ACU1 output 05 & Vista 50 Zn 6	AB	CD	AB	AB & CD
West Glassbreaks Zn 2	Glassbreak #11 interfaced with ACU1 output 05 & Vista 50 Zn 6	AB	CD	AB	AB & CD
West Glassbreaks Zn 2	Glassbreak #12 interfaced with ACU1 output 05 & Vista 50 Zn 6	AB	CD	AB	AB & CD
West Glassbreaks Zn 2	Glassbreak #13 interfaced with ACU1 output 05 & Vista 50 Zn 6	AB	CD	AB	AB & CD
West Glassbreaks Zn 2	Glassbreak #14 interfaced with ACU1 output 05 & Vista 50 Zn 6	AB	CD	AB	AB & CD

Table 11.2 Alternative Test Report

	CARD READ		FORCED			BURG RESPONSE		
DOOR NAME	Valid	Invalid	OPEN	HELD OPEN	ALARM PTS	Verified	Received	DATE
CR-101A	Y	Y	Y	Y		Y	Y	7/13/2006 & 7/25/06
CR-101B	Y	Y		Y				7/13/2006
CR-111A	Y	Y		Y				7/13/2006
CR-114A	Y	Y	Y	Y		Y	Y	7/13/2006 & 7/25/06
CR-114B	Y	Y		Y				7/13/2006
CR-115A	Y	Y		Y				7/13/2006
CR-115B(2)	Y	Y		Y				7/13/2006
CR-116A	Y	Y		Y				7/13/2006
CR-116B(2)	Y	Y		Y				7/13/2006
CR-117	Y	Y		Y				7/13/2006
CR-119A	Y	Y	Y	Y		Y	Y	7/13/2006 & 7/25/06
CR-120A	Y	Y		Y				7/13/2006
CR-120B(2)	Y	Y		Y				7/13/2006
CR-121A	Y	Y		Y				7/13/2006
CR-121B	Y	Y		Y				7/13/2006
CR-121C	Y	Y		Y				7/13/2006
CR-131A	Y	Y	Y	Y		Y	Y	7/13/2006 & 7/25/06
CR-131B	Y	Y	Y	Y		Y	Y	7/13/2006 & 7/25/06
CR-132A	Y	Y		Y				7/13/2006

Continued

Table 11.2 *Continued*

	CARD READ		FORCED			BURG RESPONSE		
DOOR NAME	Valid	Invalid	OPEN	HELD OPEN	ALARM PTS	verified	received	DATE
CR-132B	Y	Y		Y				7/13/2006
CR-132C	Y	Y		Y				7/13/2006
CR-133A	Y	Y	Y	Y		Y	Y	7/13/2006 & 7/25/06
CR-133B	Y	Y	Y	Y		Y	Y	7/13/2006 & 7/25/06
CR-134	Y	Y	Y	Y		Y	Y	7/13/2006 & 7/25/06
CR-135A	Y	Y	Y	Y		Y	Y	7/13/2006 & 7/25/06
CR-135B	Y	Y		Y				7/13/2006
CR-136	Y	Y		Y				7/13/2006
CR-137	Y	Y		Y				7/13/2006
TEMP RM 136					Y	Y	Y	7/13/2006 & 7/25/06
TEMP RM 137					Y	Y	Y	7/13/2006 & 7/25/06
GLASS BREAKS					Y	Y	Y	7/13/2006 & 7/25/06
MOTION IN CK					Y	Y	Y	7/13/2006 & 7/25/06
PANIC @ RECEP.					Y	Y	Y	7/13/2006 & 7/25/06

report but includes details such as dates and actual testing functions performed. Both of them could be acceptable to the consultant as test reports; the choice is a matter of personal preference for the consultant. Most contractors have their own test forms, but if they don't or their forms don't have all the desired information, the consultant can provide a form. The consultant needs to communicate with the contractor before testing to make sure all the elements of a test report that the consultant wishes to see are included.

MAINTAIN PUNCH LIST FOR ANY DEFICIENCIES

A *punch list* needs to be kept by someone listing all the project items left to be done or fixed by the contractor, or in some cases by the client. In theory, the person keeping the list is also the one who checks to make sure the items are complete, although that is not always the case. If the consultant is at the site frequently, then he is the person who should perform this function; otherwise, it may be the client's project manager or other designee. The consultant may also keep the list and check off the items, communicating that fact to all, with verification of completion from the client's project manager or another source.

REVIEW AND APPROVE AS-BUILT DRAWINGS AND RECORD OF COMPLETION

Along with the test report, the as-built drawings need to be reviewed and approved by the consultant before the project can be signed off as complete. As-built drawings are basically a continuation of the shop drawings as augmented through the project installation process. They demonstrate in detail exactly how the system is constructed and wired, including all cable run locations. The as-built drawings include all the shop drawings as augmented, plus floor plans with cable runs, device wiring details, and any changes to the wiring point-to-point drawings. Both the consultant and client should permanently keep a set of as-built drawings on file. If any repairs need to be done by someone other than the contractor or information about the installation is required, the as-built drawings should demonstrate to any competent professional how the system is designed and installed. As with the shop drawings, the as-built drawings need to be reviewed in detail by the consultant to make sure they are correct.

Appendix F shows an actual set of as-built drawings from a contractor for a project. They are very detailed and project specific. Any competent industry professional can look at them and know what equipment was used, exactly how it is wired, and how it functions. Although the drawings may not resemble in form to the design drawings in Appendix E, that doesn't matter. Every contractor and, for that matter, consultant will have a different style for drawings; the only thing that matters is that they are easily readable and contain all the requisite information.

A *record of completion*, sometimes called a *notice of completion* or *certificate of completion*, may also need to be filled out or written by the contractor for the client, certifying that the installation is complete and meets the specifications as written. The consultant will need to review that form and inform the client in writing if the project is actually complete. Municipal and government clients, in particular, might want a form or letter of this type. Figure 11.1 shows a sample of a notice of completion from a contractor.

APPROVE CONTRACTOR PAYMENTS

Regardless of whether the contractor is being paid by progress payments based on the percentage of work completed in a specific time period or project phase or is being paid everything at the end of the project, the consultant may need to sign off on that payment, verifying that the percentage or scope of work represented by the invoice is actually complete. This service can be offered only if the consultant is performing regular site reviews. Otherwise, the client's project manager will need to approve payments. This can be a time-consuming task with lots of forms for some municipal and government clients, so the consultant must make sure this factor is taken into account when proposing this service.

RESPOND TO OTHER REQUESTS BY CLIENT

There will inevitably be other things that the client asks the consultant to do that are technically outside the proposed scope of services. It is a judgment call by the consultant to do what is asked or not. Most requests within reason should be honored, especially if they take little time. The consultant should keep in mind that adding extra value for the client will help him greatly in the long run by adding to his credibility and enhancing potential referrals. Of course, if the request is unreasonable based on the amount of time it will take and the scope of the request, it should be treated as a change to the contract between consultant and client. Having this happen is rare; most requests are reasonable and should be honored.

The completion of the project management services means the completion of the consulting project. The completion of a successful and

Systems Integration

NOTICE OF COMPLETION

OFFICE:	New Jersey
ADDRESS:	96 Route 173 West
	Hampton NJ, 08827
PHONE:	908-735-4137
FAX:	908-735-9303

Customer Address		Job/Installation Location	
Company:	ABC Company	**Stanley Job No.**	XXXXX
Street:	1234 Main St.	**Job Name:**	ABC Company
City, State, ZIP:	Anywhere, USA 12345	**Job Location:**	Anywhere, USA
Contact Name:		**Bldg/Flr/Rm/Etc:**	BLDG # 5
Phone:		**Site Contact:**	John Smith

Reference Projects:	

System Type:
Access Control, Vista 50P Burg Panel, and CCTV

Configuration:
GE Sapphire Pro, Customer supplied server, 2 DVR's, Ademco Burg and 25 CCTV cameras.

☐ Sent Warranty Card Request to SOC ☐ Sent Customer Satisfaction Survey

Warranty Period:

Completion Date:	7/28/2006	
WORKMANSHIP – Start:	7/28/06	**Duration:** 1 Year
PARTS – Start:	7/28/06	**Duration:** 1 Year
Warranty Type/Response:	Standard business hours	

NOTE: If Service is needed before Customer receives this "Notice of Completion", the Customer may incur a charge for the Service Call unless the Customer notifies ISR Solutions within five (5) working days from the date that the Service is performed. The Parts and Labor warranty will commence on date of completion.

ISR SOLUTIONS

John R. Sottilare

Project Manager – Signature

John R. Sottilare

Project Manager – Print Name

7/26/2006

Date

CUSTOMER

Signature

Print Name Title

Date

Figure 11.1 Sample Notice of Completion

mutually rewarding project between consultant and client is a great feeling. Now is the time for the consultant to review the project with the client to make sure the client is happy with the results and to see if some things could have been done better. Constructive criticism is good. Now is also the time to ask for referral letters or testimonials. The consultant should make sure the client is happy before asking for those things.

Now it's off to the next project!

12

Forensic Consulting

A third discipline of security consulting, other than security management and security design, is called *forensic consulting*. In forensic consulting, the consultant is involved in cases of litigation ultimately by being an expert witness during a deposition or trial. This discipline also includes the review of documents, site surveys, analysis, and the writing of reports. Obviously, for forensic security consultants, the cases of litigation would have a relevant security issue that requires an expert opinion. Unlike the design consulting field in which the consultant is trying to stop an incident from occurring, in the forensic consulting field the consultant makes an assessment of the reasons that an incident has already occurred. The role of the forensic consultant is one of education, giving expert opinions, and not one of advocacy. The client of the forensic consultant is not the end user of the physical security countermeasure system, but a law firm that can represent either the defendant or plaintiff.

It is much more common for forensic consultants to come from the realm of security management consultants than from security design consultants. In fact, it is quite common for security consultants to perform the work of both management and forensic disciplines. These forensic consultants address such issues as alleged security negligence, alleged security misconduct, false arrest, and excessive use of force. Forensic consultants from the design consulting discipline address less common issues dealing with only physical security or electronic security systems. However, having a security design consultant broaden his scope of services to include forensic work is not unheard of, so this topic should be mentioned and explored in this book.

To be fair, this author does not perform any forensic work at this time and has no working first-hand knowledge of the forensic consulting field. Therefore, no details will be explained in this chapter about what services a forensic security consultant provides or any step-by-step process that the consultant needs to follow to perform those services. None of the many do's and don'ts and successes and pitfalls will be revealed. Instead, what will be described is an overview of the forensic security consulting field from the point of view of someone who doesn't offer these services but knows many consultants who do and has a general understanding of the field. This chapter will also describe the thought process that might take place for a security design consultant thinking about entering the world of forensic consulting.

PROS AND CHALLENGES

There are some clear pros and challenges (not necessarily cons) to a security design consultant entering into the world of forensic consulting. The decision should not be made lightly, as any part of the consultant's work done either well or poorly will reflect on the consultant as a whole. The consultant must be ready and capable to take on the challenge. He should keep in mind that while good forensic consulting is a matter of public record, poor forensic consulting is as well. It would be a shame if the consultant's primary business were to suffer because he wasn't fully prepared for this endeavor.

Pros

- Forensic work can be an additional source of income. This can help fill in the gaps when design work is slow, which can happen because security design consultants, for the most part, work project to project. It is always a good idea for a consultant to have multiple ways of earning income and not rely solely on one source. Writing, teaching, and speaking engagements are ways to fill in those gaps, and forensic work is a way as well.

- Forensic consulting pays well. Forensic work can be billed at 1½ to 2 times the hourly rate of regular security consulting work. In addition, in many cases forensic consultants can ask for much or most of the fee up front and get it.
- Adding expert witness experience to the consultant's résumé is an enhancement to the career and establishes the consultant as even more of an expert in the industry.
- Forensic consulting is challenging work. This is not necessarily a bad thing. Anyone who takes the plunge and has what it takes to become a security design consultant in the first place more than likely thrives on challenges and is not afraid of hard work. This is another one of those challenges.

Challenges

- There is a limited amount of forensic security work that deals only with physical security or electronic security systems. The piece of the pie may not be very big. It may also be considerably more difficult to find those types of cases of litigation or the law firms that litigate them than the normal forensic security work.
- A consultant needs a lot of experience as both a security professional and security consultant before attempting to enter the forensic consulting profession. A new consultant should not attempt to do this work. In many cases the consultant is dealing with people's lives and fortunes, so quality is of the utmost importance.
- There is another whole set of terms, rules, and standards that must be learned very well before the consultant ever begins working on a litigation case. The terms and concepts of *foreseeability*, *duty*, *proximate cause*, and *torts* are just a few. The Federal Rules of Civil Procedure—Rule 26 Discovery and Depositions and the Daubert Rule must be known and understood. This will take time and effort. The consultant should consider forensic security consulting to be a whole new business that must be learned from the beginning.
- Forensic consulting is hard, challenging work. Attorneys are generally a nightmare to deal with, and being cross-examined by one either in deposition or at trial is no picnic. Forensic consultants laugh with each other after the fact, telling horror stories about experiences with attorneys. There is a lot of tedious reading of material in which details cannot be missed. There's a reason this work pays well: it's not easy.
- There is not as much flexibility with scheduling of the consultant's work for forensic consulting as for design consulting. Trial dates are what they are, depositions need to take place by certain times,

and material must be read and responded to by certain times, regardless of the consultant's schedule.

If a security design consultant really feels he is ready and capable of taking on this challenge and is serious about putting in the time and effort to be prepared to do quality work, by all means he should do so.

GETTING STARTED

If a security design consultant wants to take on the challenge of adding forensic work to his scope of available services, the following sections provide a few suggestions for first steps.

Find a Mentor

Most forensic security consultants started off knowing close to nothing. They learned not only from research and trial and error but from picking the brains of other consultants doing the same thing. This is the reason that belonging to an association like the International Association of Professional Security Consultants (IAPSC) is so important. There are also associations specifically tailored to expert witnesses. These associations give seminars on the subject and allow for one-on-one communication with peers. There are a lot of people out there willing to help others succeed in their field because mentoring is a quality of a true professional.

Find an Attorney as an Advocate

Everyone knows an attorney, for better or worse. The consultant should find one who can help with some tips on how to break into the forensic consulting field. Attorneys are always looking for expert witnesses, and they keep files on potential ones. An attorney might be able to help that process along. Even better, the consultant should try to get that attorney to refer another attorney who deals with litigation cases involving some type of physical security or electronic security systems, such as a law firm that represents security systems manufacturers. Because security design consulting is a specialized field in comparison to the services provided by management consultants, the litigation cases design consultants are expertly qualified to work on are limited and probably so are the law firms handling those types of litigation cases.

Research to Obtain Education

As with any professional endeavor, everything possible should be researched and read about the discipline. The consultant must be educated about the forensic consulting field. There are excellent books about forensic consulting available through the publisher of this book and through the American Society for Industrial Security (ASIS), among many other sources. As mentioned previously, many associations hold seminars on the subject and allow for group and one-on-one communication with peers in the industry. Actually attending a trial or two where a forensic security consultant is testifying would be a great education. There are a million other ways to obtain an education on a subject in today's world.

Prepare a CV

A curriculum vitae (CV) is like a résumé, except much more detailed. It includes employment history, teaching history, education, literary and speaking contributions, professional affiliations and accomplishments, certifications, training, etc. It must be inclusive of everything performed and accomplished in the consultant's career with nothing left out. This is what attorneys collect and have on hand, as well as marketing material, for when they need to find an expert witness. Everyone will want a copy of the consultant's curriculum vitae. Therefore, the consultant will need to write and review one.

Market

Just as with the rest of the security consulting business, expert witness work must be marketed. The target market is attorneys who handle these types of litigation cases. There are legal publications that are a good place to advertise for expert witness work. Some are more expensive than others, and some have a wider readership than others, so the consultant should check with his new attorney friend or mentor for what might get the best results. There is no reason the process of marketing should be any different for forensic consulting work as for the design consulting work; it should just be adapted for the new marketplace.

BEST PRACTICES

The International Association of Professional Security Consultants (IAPSC) has published a *Best Practice Guideline for Forensic Methodology*. Any consultant wanting to delve into forensic work should read and learn this methodology as part of his initial education as well as a refresher along the way. Without going into any details, for which this author is not qualified, the methodology breaks down the process of the forensic projects into five elements. The tasks within these elements all are performed before any deposition or trial. The five elements include

1. *Risk Assessment:* Reviewing all the material relevant to the litigation case
2. *Security Survey:* Visiting and reviewing the scene of the incident
3. *Analysis:* Determining the level of adequacy of security
4. *Conclusion:* Reaching conclusions on the issues of foreseeability, preventability, and causation
5. *Report:* Writing a report with the consultant's opinions

The best practices document details all the tasks within each of these five elements that should be followed for a typical premise security case. These best practices can be found at www.iapsc.org.

ETHICS

As with the other security consulting disciplines, and maybe even more so, there must be ethical standards that are followed by members of the profession. Here is the remaining portion of the Code of Ethics published by the International Association of Professional Security Consultants (IAPSC):

Forensic Consulting

1. Consultant's fees will never be contingent upon the outcome of a case.
2. Consultants, when testifying, will carefully avoid taking the position of an advocate or appearing to take such a position; for justice requires the professional expert witness to be neutral with no personal interest in the outcome of the case.
3. If, after reviewing a case, it is apparent that the expert witness cannot provide testimony or assistance helpful to the case, the consultant will make this known to the client. If he withdraws from or his services are discontinued from the case, he will not testify for the opposing side unless compelled to by subpoena.
4. The consultant will not sign written opinions or affidavits prepared by clients. Testimony or report preparation, including the preparation of oral reports, will not occur until the consultant has performed a thorough evaluation of the circumstances, evidence, scene or other pertinent materials or places as he deems necessary to render a learned opinion.

The hope is that this information helps any security design consultant thinking about taking on the challenge of adding forensic work to his list of provided services. By no means should that idea be dismissed out of hand, but it must be thought out very carefully to ensure the resulting work is of the quality expected by the industry, the client, and, most importantly, the consultant as well.

13

Continuing Education

Professionals in every field and endeavor require continuing education to keep up with the changes within their industries. The security industry in general and security design consulting in particular are no exception. The security design consultant has to continue that education in two fields: the technical security field and the consulting field. Those two fields combine to make up the discipline of the professional security design consultant.

Some of the ways the security design consultant can continue this education is by joining professional organizations, obtaining and keeping professional certifications, attending professional seminars, and subscribing to publications. This chapter will describe some of these efforts. The lists within this chapter are by no means inclusive. Individual consultants, based on specialty, may find it necessary and appropriate to take advantage of different avenues of continuing education than those listed here. By all means, they should do so. The more educated about the profession a consultant is, the better end product will be produced for the client, and therefore the more successful the consultant.

PROFESSIONAL ASSOCIATIONS

There are three main professional associations in which membership is appropriate and necessary for a security design consultant.

International Association of Professional Security Consultants

The International Association of Professional Security Consultants (IAPSC) is the premier and most respected consulting association in the security industry. The organization has been mentioned many times throughout the body of this book. The IAPSC has three primary purposes. The first is to offer potential clients a single source to find professional, independent security consultants. No members of the IAPSC sell security equipment or guard services or receive any compensation for any recommendations that may be made. The second is to provide a forum for those professional consultants to obtain professional growth and network with their peers. The IAPSC holds an annual conference including educational sessions and networking opportunities. In addition, the IAPSC website has a forum for members to share ideas and information. The third is to promote the security consulting field by establishing and maintaining the highest set of standards for professionalism and ethical conduct in the industry.

Twice a year, just before the IAPSC and ASIS conventions, the IAPSC also holds a Successful Security Consulting class that lasts two days. Industry professionals who are also members of the IAPSC make up a faculty that teaches both new security consultants and those thinking of becoming security consultants real-world concepts to help start and maintain successful security consulting businesses. This author is a former student at one of these classes and can personally attest that this is the best education a beginning security consultant can possibly obtain. The fact that the IAPSC in general and these consultants that make up the faculty specifically are willing to share their knowledge in this forum is a testament to the quality and professionalism of the IAPSC and its members.

The membership of the IAPSC includes security management consultants, security design consultants, and forensic security consultants. The association has stringent membership requirements including education, experience, and proof of independence. This author believes that the IAPSC is the most important association for a security design consultant to belong to and every new consultant should strive to meet the membership requirements and join the IAPSC.

Contact Information:
IAPSC
525 SW 5th St., Suite A
Des Moines, IA 50309-4501
515-282-8192
www.iapsc.org
iapsc@iapsc.org

American Society for Industrial Security

The American Society for Industrial Security (ASIS) International is the largest international organization (over 34,000 members) for professionals responsible for security. Its members include managers and directors of security; corporate executives; consultants; architects; attorneys; and federal, state, and local law enforcement officials. ASIS is dedicated to increasing the effectiveness and productivity of security practices by developing educational programs and materials that address broad security concerns. These include an annual conference, a monthly magazine, bookstore, and on-line and regional seminars.

In addition, ASIS has local chapters that allow for networking opportunities with local peers and more educational opportunities. Because keeping up with the new security technology is a vital part of this career, every security design consultant should be a member of ASIS.

Contact Information:
ASIS International
1625 Prince St.
Alexandria, VA 22314-2818
703-519-6200
www.asisonline.org
asis@asisonline.org

National Fire Protection Association

The National Fire Protection Association (NFPA) is an international nonprofit association whose mission is to reduce the worldwide burden of fire and other hazards on the quality of life by developing and advocating consensus codes and standards, research, training, and education. Its over 79,000 members come from all aspects of the fire protection industry. The NFPA offers professional development to include an international conference, seminars, expos, and access to all NFPA codes and standards. Any security design consultant who offers fire system design as part of his services should belong to the NFPA.

Contact Information:
NFPA
1 Batterymarch Park
Quincy, MA 02169-7471
617-770-3000
www.nfpa.org

Other Associations

Security design consultants who specialize in certain industries may also find it necessary and helpful to join the professional association pertinent to that industry. It will allow for networking and specific industry contacts as well as continuing education about the industry and how the services provided by the consultants can be helpful. Consultants will need to research to find the appropriate professional association for their specialties, but here is a list of just a few:

Health Care:
International Association for Healthcare Security & Safety (IAHSS)
www.iahss.org
National Association of Healthcare Consultants
www.healthcon.org

Building Management:
Building Owners and Managers Association (BOMA)
www.boma.org
Institute of Real Estate Management (IREM)
www.irem.org

Education:
National Association School Resource Officers
www.nasro.org
Association of School Business Officials International
www.asbointl.org

There are also associations dedicated to the consulting profession in general that may be of interest to security consultants. A couple of them are

Institute of Management Consultants
www.imcusa.org
American Association of Professional Consultants
www.consultapc.org

CERTIFICATIONS

Professional certifications are a measure of a person's competency in his chosen profession. As described in the first chapter of this book, security design consultants should hold the appropriate certifications as a statement of their professional excellence. Certifications are not just a representation of the consultant's knowledge at the time of the certification acquisition, but of his continuing knowledge and industry involvement through the years. Good certifications require a recertification process every few years in order for the person to continue to hold the certification. The criteria for recertification include continuing education, contributions to the industry, and participation in industry events.

This author feels a security design consultant should acquire and continue to hold a minimum of two certifications: either the Physical Security Professional (PSP) or Certified Protection Professional (CPP) (holding both is a bonus) and the Certified Security Consultant (CSC).

Physical Security Professional

The Physical Security Professional (PSP) is a certification sponsored and administered by the American Society for Industrial Security (ASIS). This certification is designed for those whose primary responsibility is to conduct threat surveys; design integrated security systems that include equipment, procedures, and people; or install, operate, and maintain those systems. It is the perfect certification for the technical aspect of the security design consulting profession. The subjects tested for the Physical Security Professional designation as described on the Certification section of the ASIS website (www.asisonline.org) include those described in the following sections.

Physical Security Assessment

- Identifying assets to determine their value and criticality
- Assessing the nature of the threats so that the scope of the problem can be determined
- Conducting a physical security survey in order to identify the vulnerabilities of the organization
- Performing a risk analysis so that appropriate countermeasures can be developed

Selection of Integrated Physical Security Measures

- Identifying measures/components to match the requirements of the appropriate solution/recommendation

- Performing cost analysis of the proposed integrated measures to ensure efficiency of implementation/operation
- Outlining/documenting recommendations with relevant reasons for presentation to a facility so that appropriate choices can be made

Implementation of Physical Security Measures

- Outlining criteria for prebid meetings to ensure comprehensiveness and appropriateness of implementations
- Procuring systems and implementing recommended solutions to solve identified problems
- Conducting final acceptance testing and implementing/providing procedures for ongoing monitoring and evaluation of the measures

Certified Protection Professional

The Certified Protection Professional (CPP) certification is also sponsored and administered by the American Society for Industrial Security (ASIS). This certification is designed for individuals who have demonstrated competency in all areas constituting security management. The CPP is the most recognized certification for security professionals and was the only certification available to security design consultants prior to the advent of the PSP. Notice that only a small percentage of the subjects tested for a CPP certification apply to physical security and the security design consultant. The subjects tested for the Certified Protection Professional certification as described on the certification section of the ASIS website (www.asisonline.org) include those described in the following sections.

Security Principles & Practices

- Plan, organize, direct, and manage the organization's security program to avoid and/or control losses and apply the processes necessary to provide a secure work environment
- Develop, manage, or conduct threat/vulnerability analyses to determine the probable frequency and severity of natural and man-made disasters and criminal activity on the organization's profitability and/or ability to deliver products/services
- Evaluate methods to improve security and loss prevention systems on a continuous basis through the use of auditing, review, and assessment

- Develop and manage external relations programs with public sector law enforcement or other external organizations to assist in the achievement of loss prevention objectives
- Develop and present employee security awareness programs to achieve organizational goals and objectives

Business Principles & Practices

- Develop and manage budgets and financial controls to achieve fiscal responsibility
- Develop, implement, and manage policies, procedures, plans, and directives to achieve organizational objectives
- Develop procedures/techniques to measure and improve organizational productivity
- Develop, implement, and manage staffing, leadership, training, and management programs in order to achieve organizational objectives

Personnel Security

- Develop, implement, and manage background investigations in coordination with other departments and agencies for the purpose of identifying individuals for hiring and/or promotion
- Develop, implement, manage, and evaluate policies, procedures, programs, and methods for personnel protection (excluding executive protection) to provide a secure work environment
- Develop, implement, and manage executive protection programs to reduce security risks to executives and to ensure continued viability of the organization
- Support the organization's efforts to reduce substance abuse in the workplace

Physical Security

- Survey facilities in order to manage and/or evaluate the current status of the physical security, fire detection and emergency and/or restoration capabilities
- Select, design, implement, and manage security measures to reduce the risk of loss
- Access the effectiveness of the security measures by testing and monitoring

Information Security

- Survey information facilities, processes, and systems to evaluate current status of physical security, procedural security, information systems security, employee awareness, and information destruction and recovery capabilities

- Develop and implement policies and standards to ensure information is evaluated and protected against all forms of unauthorized/inadvertent access, use, disclosure, modification, destruction, or denial
- Develop and manage a program of integrated security controls and safeguards to ensure confidentiality, integrity, availability, authentication, non-repudiation, accountability, recoverability, and audit ability of sensitive information and associated information technology resources and assets
- Evaluate the effectiveness of the information security program's integrated security controls to include related policies, procedures, and plans to ensure consistency with organization strategy, goals, and objectives

Emergency Practices

- Mitigate potential consequences of emergency situations by identifying and prioritizing potential hazards and risks and developing plans to manage exposure to loss
- Prepare and plan how the organization will respond to an emergency
- Manage the activation of the emergency response plan to reduce loss
- Recover from emergency situations through management of the restoration of vital services and facilities to minimum standards of operations and safety

Investigations

- Develop and manage investigative programs
- Manage or conduct the collection and preservation of evidence to support post-investigation actions (employee discipline, criminal or civil proceedings, arbitration)
- Manage or conduct surveillance processes
- Manage or conduct investigative interviews

Legal Aspects

- Develop and maintain security policies, procedures, and practices which comply with relevant elements of criminal civil, administrative, and regulatory law to minimize adverse legal consequences
- Provide coordination, assistance, and evidence such as documentation and testimony to support legal counsel in actual or potential criminal and/or civil proceedings

- Provide advice and assistance to management and others in developing performance requirements and contractual terms for security vendors/suppliers and establish effective monitoring processes to ensure that organizational needs and contractual requirements are being met
- Provide assistance to management, legal counsel, and human resources in developing strategic and tactical plans for responding to labor disputes, including strikes
- Develop and maintain security policies, procedures, and practices that comply with relevant laws regarding personnel security
- Develop and maintain security policies, procedures, and practices that comply with relevant laws regarding information security

Certified Security Consultant

The Certified Security Consultant (CSC) certification is sponsored and administered by the International Association of Professional Security Consultants (IAPSC). This certification is specifically designed for the security consultant and is currently the only certification designed for that profession. It is the perfect certification for both the consulting and technical aspects of the security design consulting profession. The subjects tested for the Certified Security Consultant certification as described on the certification section of the IAPSC website (www.iapsc.org) include those described in the following sections.

Consulting

- Understand basics of consulting business
- Perform functions needed for securing clients
- Understand the phases of proposal writing
- Implement work product

Security Consulting Methods

- Understanding management security consulting
- Understanding technical security consulting
- Understanding forensic consulting

Business Ethics

- General ethical standards

There may be other certifications pertinent to the consultant's specialty that would be helpful to obtain. The consultant should check with the professional association appropriate for that specialty to determine what certifications are available and their impact within the industry.

SEMINARS AND CONFERENCES

Attending conferences, seminars, and expositions is a must for a security design consultant. Not only is current product information acquired at these events, as described in Chapter 8, but they offer educational sessions on all types of security topics and provide an unmatched networking opportunity to talk to and learn from peers in the industry. Time constraints limit the number of conferences that can be attended in any one year, but attending two a year is a reasonable expectation. Some of the more well known of these events include

1. ASIS International Seminar and Exposition
2. International Security Conference (ISC) sponsored by the Security Industry Association (SIA)
3. NFPA World Safety Conference and Exposition
4. IAPSC Annual Conference

Professional associations pertinent to a consultant's specialties or dedicated to consulting itself may also have seminars or conferences that are worth attending.

PUBLICATIONS

Numerous security publications are available to the security design consultant, more than he could possibly read on a monthly basis and still have time to actually perform some work. However, reading a few of them on a regular basis is helpful to keep up on changing technology, new products, and industry news. The consultant should choose a few of them and not subscribe to the rest. Some of these publications include

1. *Security Management* magazine from ASIS
2. *Security Products*, www.secprodonline.com
3. *Security Technology & Design* from Cygnus Business Media
4. *Security Sales & Integration* from Bobit Business Media
5. *Consulting–Specifying Engineer* from Reed Business Information
6. *NFPA Journal* from NFPA

Again, publications within a consultant's specialty industry may be worthwhile to subscribe to and read.

Appendix A
Sample Proposals

This appendix includes two actual proposals that were sent to potential clients. The first is from a security design consulting firm, and the second is from a security management consulting firm where the design consultant partnered with the management consultant on the project.

STRATEGIC DESIGN SERVICES, LLC

SECURITY ASSESSMENT • SYSTEM DESIGN • PROJECT MANAGEMENT

FIRE / BURGLARY / CCTV / ACCESS CONTROL / GATE OPERATORS

SECURITY CONSULTING PROPOSAL

ABC COMPANY
1224 MAIN ST.
ANYWHERE, USA 12345

I SUMMARY

Strategic Design Services, LLC proposes to provide assessment, design and project management services for an IP based CCTV system installation project at the ABC Company. The overall objective of the project is

- Perform as assessment of the resort as it pertains to designing an IP based CCTV system based on the needs of the facility.
- Prepare complete design specifications and drawings for the CCTV system so they are ready for an Invitation For Bid.
- Provide project management services during the procurement and construction phases of the CCTV system installation project.

The approach to this proposal and to the project itself is to give the most value in terms of deliverables, at a reasonable price. Additionally, we want the end product to be of unquestionable quality; therefore we will employ both industry standards and best practices for the design.

What makes SDS an excellent choice for this project is our total independence from the installation field, the certification and background of the company's principal, the clarity and detail of our scope of services, and our determination to produce a quality work product.

The project will be conducted in three phases as described in the Scope of Services section.

II PROJECT APPROACH

Strategic Design Services, LLC is an independent security consulting firm. We are not affiliated in any way with security guard services, electronics installation, hardware, or software providers, and we do not accept any finder's fees or commissions with respect to our recommendations. Therefore, our counsel is objective, and recommendations are presented only on the basis of the ABC

P.O. Box 436 • Portland, CT 06480 • (860) 342-2544 • Fax (860) 342-0542 • strategicdesignservices.com

Company's needs, rather than any outside interests. We do not have the inherent conflict of interest that companies do who both consult and install.

Throughout this project you can be assured of professional and responsive service of the highest integrity. We adhere to the Codes of Conduct and Ethics of the International Association of Professional Security Consultants (IAPSC), and the American Society for Industrial Security (ASIS).

SDS's principal, Brian Gouin, PSP, CSC is an active member of the International Association of Professional Security Consultants. The International Association of Professional Security Consultants, Inc. (IAPSC) is the most respected and widely recognized consulting association in the security industry. Its rigid membership requirements ensure that potential clients may select from the most elite group of professional, ethical and competent security consultants available to them.

The primary purpose of the IAPSC is to establish and maintain the highest set of standards for professionalism and ethical conduct in the industry. The members are independent of affiliation with any product or service they may recommend in the course of an engagement, thus ensuring that the services they render are in the best interests of the client.

III CONSULTANT BACKGROUND

Brian Gouin, PSP, CSC

Strategic Design Services LLC's principal, Brian Gouin, has over 17 years of experience in the security and fire protection field. He has extensive experience in electronic security hardware and software, industry installation best practices, and overall physical security. Since 2001 he has been the owner and principal security consultant for a firm specializing in assessment, system design, and project management for electronic fire, burglary, CCTV, access control, gate operator and monitoring facility systems for all industries. Previous to that, for 12 years Brian owned a successful security installation company serving hundreds of clients which designed, sold, and installed commercial, institutional, industrial, and residential burglary, fire, CCTV, access control, and gate operator systems. The company also actively participated in the design, installation, and review of many municipal monitoring station facilities.

Brian's background in the installation field is a tremendous asset as a consultant. His experience is that of the real world and not just theory. His knowledge of the assessment, design, procurement, and construction phases of a project from all points of view helps to make the whole project go smoother and ultimately helps to achieve a better quality end product.

Additionally, Brian is Board Certified in Physical Security, having achieved the distinguished "Physical Security Professional" (PSP) designation. The PSP is awarded based upon experience, education, and passage of an examination that provides an objective measure of an individual's broad based knowledge and competency in physical security. Ongoing professional development is required in order to maintain the credential. The PSP is administered by ASIS International.

Brian has also achieved the "Certified Security Consultant" designation. The CSC is awarded based on experience, education, and passage of an examination of an individual's knowledge of general consulting, security consulting disciplines, and business ethics. The CSC examination is administered by the IAPSC.

Brian has extensive training in system design from a vast number of manufacturers of electronic security and fire equipment. His professional affiliations include the International Association of Professional Security Consultants (IAPSC), the American Society for Industrial Security (ASIS), the National Fire Protection Association (NFPA), the Building Fire Safety Systems Section of NFPA, and the National Association of Chiefs of Police.

Brian is currently security cleared by the Federal Protective Service (part of DHS) to perform Quality Control audits of security personnel at Federal facilities.

IV SCOPE OF SERVICES

PHASE 1: Assessment

Part 1: Site Visit

The site visit will consist of a walk through of the facility and interviews with ABC Company personnel to determine the issues concerning the facility and the areas that require CCTV surveillance. Determinations will be made as to best camera placement based on required coverage area, power availability, network connection availability, ease of installation and maximum overall effectiveness. A night time observation will also be made to determine if IR illumination is required. In addition, the possibility of wireless IP cameras and video image printing will be explored and researched. Required personnel will be interviewed to determine budget issues and project schedule.

Part 2: Design

An IP based CCTV system will be designed using our expertise, industry best practices, the project budget and the information gathered from the site visit. The system shall include approximately 30 IP color cameras, a digital recorder with 30 days storage and other related network and web based equipment necessary for a completely operational CCTV system. The equipment choices will be based totally on the needs of the facility including expandability concerns.

Part 3: Report

A brief report will be written for the ABC Company outlining all the elements of the designed IP based CCTV system including specific manufacturers and model numbers, device locations and head end equipment capabilities. An explanation will be given for all product choices. An estimated system installation cost will also be provided. There will be no sole source products chosen. The purpose of the report is to get ABC Company approval of the designed system before specifications are written. If any alterations of the designed system are required, those adjustments will be made.

PHASE 2: Design Specifications

1. Comprehensive design and technical specifications for the designed CCTV system will be prepared with sections including but not limited to:
   - A. Description and scope of work
   - B. Necessary contractor qualifications
   - C. Material descriptions
   - D. A&E specifications for all products
   - E. Installation and testing requirements
   - F. Execution information.

The specifications will be detailed and comprehensive so there is no misunderstanding as to what is required for the installation of the designed system. They are intended to be ready for an Invitation For Bid with the addition of the ABC Company's standard bidding documents.

2. Drawings will be prepared to include:
   - A. Site plan showing location of all devices and head end equipment
   - B. Detail drawing
   - C. Riser diagram

PHASE 3: Project Management

The following project management services will be provided during the procurement and construction phases of the project:

1. Assistance in locating and inviting qualified Contractors to bid.
2. Leading the pre-bid conference and walk through.
3. Provide written answers to questions at pre-bid walk through.
4. Assistance in evaluation of bids, including Contractor qualification review.

5. Response to any Contractor Requests For Information (RFI).
6. Check, review, and approve installation progress and techniques. This includes a minimum of three visits, more if necessary for a proper review.
7. Review and approve any change orders or changes in scope or Work requirements.
8. Witness test of system with Contractor.
9. Approve Contractor progress payments.
10. Any other request of the ABC Company deemed reasonable by Strategic Design Services LLC.

V ENGAGEMENT CONSIDERATIONS

It will be necessary for the ABC Company personnel to fully support this project, working together with the consultant to resolve any issues as a result of the engagement. The ABC Company agrees that neither party can unilaterally achieve successful results by having the project completed on time and within the agreed upon cost without cooperation from both consultant and client. Therefore in connection with this project the ABC Company agrees that:

1. One person will be designated as the ABC Company point of contact and will act as facilitator for SDS during the course of the engagement and assist in gaining access to necessary information and answers to questions.
2. The ABC Company will make available any drawings of the site and any buildings on the site. If any of these drawings are available in AutoCAD, they will be provided.
3. SDS will have unfettered day and night time access to the required areas of the facility during the site visit.

Timeline

It is anticipated that after receiving your authorization to proceed, this project can start according to an agreed-upon timeline. The assessment report should be completed within one month and design specifications within two months after your authorization to proceed.

Project Fee

We believe that the appropriate method to evaluate the investment in security consulting services is the total investment for any given engagement from start to conclusion compared against the anticipated outcome that the client desires. Therefore, we propose projects such as this on a fixed project fee basis. This provides the ABC Company with a firm fixed price-up-front, for all work to be done.

Accordingly, the price to the ABC Company for all work proposed herein is $8,988, which will be billed on a project phase basis. The breakdown in cost by phase is: Phase 1: $3,180, Phase 2: $2,948, Phase 3: $2,860.

VI CHANGES IN PROJECT SCOPE

The services outlined herein are based on our understanding of your needs and are subject to your approval and revision. Should you wish to change, delete or add any areas of the design, we will be pleased to provide an estimate of any additional costs or reductions of cost from this proposal.

VII CONFIDENTIALITY

Due to the sensitive nature of security consulting projects, all data collected, our analysis of the data, any recommendations to the ABC Company resulting from our analysis, including any other information not then in the public domain received from the ABC Company or its employees, <u>including the fact or existence of the consultation itself</u>, will be held in strict confidence by us to the extent permitted by law, unless the ABC Company agrees otherwise.

In accordance with the above mentioned confidentiality statement, our policy on client identification is clear: except for projects completed for government agencies which are a matter of public record, we will not reveal the identity of a client of our organization without prior written approval; not for referrals, advertising, nor for any other reason unless required by legal proceedings, administrative agency, or other authority with proper jurisdiction. Our clients appreciate this, and many require this approach.

Submitted August 21, 2006,

Brian Gouin, PSP, CSC

CODE OF ETHICS

Strategic Design Services, LLC assures its clients and prospective clients that they will be provided with professional and responsive service of the highest integrity. Strategic Design Services, LLC will:

I Accept only assignments for which we are qualified, and charge no more than a reasonable fee.

II Exercise independence of thought and action with regard to clients and potential suppliers and, at all times, insure that our advice and recommendations are based upon impartial consideration of all known pertinent facts.

III Neither accept fees or commissions for referring or recommending the services or products of other persons or firms to prospective clients.

IV Perform our professional duties in accordance with the Law.

V Not take personal, financial, or any advantage of information gained by virtue of a professional-client relationship.

VI Handle all information concerning the affairs of clients as Confidential.

VII Maintain the highest standards of professionalism, honesty, and integrity.

VIII Place the interests of our clients ahead of our own, and serve clients with competence at all times.

Document A-1 Proposal from Security Design Consulting Firm

August 4, 2006

John Smith, Security Manager
ABC Hospital
1234 Main St.
Anywhere, USA 12345

Re: **Proposal for Security Consulting Services #TDD-06004**

Dear Mr. Smith:

Thank you for the opportunity to provide ABC Hospital with a proposal for security consulting services. Threat Analysis Group, LLC (TAG) is an independent security consulting firm specializing in risk management and security. TAG's distinguished clients include Fortune 500 companies and government agencies. We look forward to adding ABC Hospital to our list of satisfied clients.

Careful consideration has been given to each of the tasks needed to complete the security consulting project and TAG is well suited and qualified to perform each. As you will see, TAG is poised to provide ABC Hospital with a highly attractive cost to benefit ratio.

I look forward to working with you. If you have any questions or concerns, please contact me at your convenience.

Sincerely,

Karim H. Vellani, CPP, CSC
Senior Consultant

P.O. Box 16640 ▪ Sugar Land, TX 77496 ▪ (281) 494-1515

ABC Hospital
Security Consulting Services

Project Objectives

The objectives of this project are to:

- Identify threats (real and perceived)
- Identify procedural and physical vulnerabilities
- Prioritize security risks
- Ensure appropriateness of security policies and procedures
- Evaluate effectiveness of the ABC Security Personnel
- Evaluate effectiveness of Physical Security Systems, including Access Control and CCTV
- Define and prioritize security measures and achieve appropriate layering of security features and functions to ensure overall security and safety of the hospital, staff, patients, visitors and critical assets

Project Scope

The project includes an assessment of the total security program for ABC Hospital including the primary hospital building, the adjacent physical therapy building, and the Amesbury building. The scope addresses a total security assessment including personnel, electronic and physical security controls, and policies, procedures, and practices. After reviewing the identified buildings, observing the guard force, and interviewing key stakeholders, we will provide specific recommendations for security enhancements based on identified threats and existing vulnerabilities. Specifically, this will include the following:

- Assess and evaluate ABC Hospital's overall security program for adequacy of:
  - — Security policies, procedures, and practices
  - — Security force staffing levels and usage
  - — Physical security components and their effectiveness (CCTV and Access Control Systems)

- Assess and evaluate community crime impact in the following areas:
  - Historical crime and security threats to the facilities
  - Conceptual threats (natural disasters, terrorism)
  - Crime/security sources near the facilities
- Assess and evaluate the ABC Hospital Security Force:
  - Current methods of deployment
  - Functions and activities
  - Turnover reduction strategies
  - Increasing effectiveness

Project Methodology

TAG utilizes a performance-based, industry specific risk assessment methodology that incorporates effective deter, detect, delay and response criteria for protecting critical assets. Our consultants take a holistic approach to security program evaluation based on our own experience in security management, technical security, and the development and enhancement of hundreds of security programs for our clients. Following is a summary outline of the risk assessment methodology TAG will employ for ABC Hospital:

PHASE 1: PRE SITE VISIT

1. Schedule interviews with appropriate project stakeholders.

   Interviews will be conducted with the appropriate project stakeholders, including management and staff members. The overall objective of the interviews is to identify critical assets and determine the specific areas of concern from a security point of view and to gather information regarding current personnel, policies and procedures. At a minimum, the following people will be interviewed:

   - Emergency Department Head
   - Quality Risk Department Head
   - Birth Center Department Head
   - Psychiatric Department Head
   - Admitting Department Head
   - Other Department Heads and/or Staff Members as appropriate

2. Threat Assessment

CrimeAnalysis™: Obtain and review incident and crime reports from the ABC Hospital Security Department and local and state law enforcement agencies. TAG will utilize our CrimeAnalysis™ software application to identify crime threats to the identified facilities. As a critical component of a risk assessment, CrimeAnalysis™ allows for the comprehensive evaluation of a community's security posture while working within budgetary restrictions. Specifically, it helps security decision makers select appropriate security measures, manage vulnerabilities, and document due diligence. The Federal Bureau of Investigation's Uniform Crime Report definitions and coding system are employed in the software. TAG's methodology is peer-reviewed, court-proven, and published in the book, Applied Crime Analysis. CrimeAnalysis™ software and/or report's key benefits include:

— Objective Threat Identification – Uses crime data obtained from law enforcement agencies to monitor threats to the hospital.

— Crime Trending – Analysis of existing and emerging threat patterns

— Threat Typologies – Determine precise nature of each threat

— Security Decision Validation – Using crime data to drive a security program allows ABC Hospital to justify security decisions and save money

— Security Personnel Deployment – Helps efficiently deploy security perso nnel when threats are high (day of week, time of day, etc.) and reduce security personnel coverage when threats decrease

— Return on Investment – Obtain a measurable return on investment through cost savings on security personnel deployment, effective implementation of physical security measures, and reduced liability exposure.

Internal Reporting: Daily Activity Reports & Security Incident Reports

Other Threat Information Sources:
— Hospital/Healthcare Trends
— Community Crime Generators
— Law Enforcement Liaison
— Terrorism

- Natural Disasters (Weather)
- Pandemics

PHASE 2: SITE VISIT

Part 1: Site Review

- Follow up interviews with stakeholders
- Daytime and nighttime observations
- Review of existing security policies, procedures and practices
- A security review of any areas of concern brought out in the interviews to determine what, if any, security enhancements can or should be installed to help deter and prevent security problems
- Special Characteristics of each Facility
- Security Incident Reports Review

Part 2: Physical Security Assessment

- An assessment will be performed and any necessary recommendations will be made regarding the following physical security issues:
    - Lock and key control
    - Perimeter barriers and controls
    - Vendor and delivery access
    - Perimeter entry point access
- Interior and exterior lighting will be evaluated in detail:
    - The lighting will be evaluated using the *Guideline for Security Lighting for People, Property, and Public Spaces* (IESNA G-1-03 from the Illuminating Engineering Society of North America)
    - The security lighting principles that will be utilized include:
        - Integration of illumination into the total security system, thereby facilitating the effectiveness of other security devices or procedures

        - Illumination of objects, people, and places to allow observation and identification, thereby reducing criminal concealment

        - Illumination to deter criminal acts by increasing fear of detection, identification, and apprehension

        - Lessening the fear of crime by enhancing a perception of security

        - Illumination that allows persons to more easily avoid threats, and to take defensive action when threats are perceived

    - Based on the above guidelines, determinations will be made on the need for additional lighting.

- Electronic access control systems will be evaluated in detail:

    - Evaluate the functionality of the current Software House C·CURE access control system, including expandability.

    - Based on the information gathered from interviews, industry best practice and our expertise, determinations will be made on the need for additional access control devices.

    - Recommendations will be made on how to best implement any recommended access control devices, including rough cost estimates.

- Electronic Closed Circuit Television (CCTV) systems will be evaluated in detail:

    - Evaluate the functionality of the current American Dynamics Intellix Digital Video Management System and associated cameras, including expandability.

    - Based on the information gathered from interviews, industry best practice and our expertise, determinations will be made on the need for additional CCTV devices.

    - Recommendations will be made on how to best implement any recommended CCTV devices, including rough cost estimates.

Part 3: Security Force Assessment

TAG consultants will observe Security personnel to observe and ask questions. The objectives of this observation include:

- Determine level of security knowledge and ascertain areas for improvement through training
- Identify security functions and activities and areas for improvement
- Identify causes of turnover and potential solutions
- Determine the optimum placement of electronic security tour checkpoints (if recommended) from a functional perspective

PHASE 3: RECOMMENDATIONS

Based on our expertise, industry standards and best practices, information gathered from the interviews, our Assessment of the Security Force and Physical Security Systems, the following will be provided:

1. Risk Matrix, Mitigation Strategies, & Recommendations
   - Identified threats and suggested countermeasures
   - Prioritized Recommendations tied to Specific Threats/ Vulnerabilities
   - Objective is to find appropriate solutions to security vulnerabilities, taking full advantage of existing resources
   - Compliance with Industry Standards, Best Practices, and Benchmarking
     - JCAHO
     - ASIS Security Risk Assessment Guideline
     - IESNA Lighting Standards
2. Recommendations for physical security enhancements to increase the security and safety of the affected facilities and critical assets.
   - Reducing crimes and security issues though the deployment of electronic security measures, such as cameras and electronic tour checkpoints
   - Alternative strategies to the recommended security enhancements

3. Recommendations for enhancing Guard force effectiveness and efficiency.

   — Increasing professionalism through measures such as training, supervision, and/or certification

   — Cost estimates for any recommendations relating to the Security Force

4. Recommendations for security policies, procedures, and practices which will support the Physical Security and Security Force aspects of the security program.

PHASE 4: DELIVERABLES

1. A report will be written to include all the information gathered from the project with a complete explanation of all recommendations. Five (5) copies of this report will be delivered to the ABC Hospital Security Manager.

2. CrimeAnalysis™ software with an open license to install on any ABC Hospital computer. If preferred, CrimeAnalysis™ hard-copy reports will be provided in lieu of the software.

Project Team Qualifications & Experience

TAG's ABC Hospital Project Team has been configured to ensure that the unique needs of ABC Hospital are addressed. The Project Team has extensive experience with hospital security, physical security systems, and security personnel management, all of which are relevant to the project at hand. All of our consultants are highly trained, certified in security, and bring unparalleled healthcare security experience to this project. The team's principals have extensive experience in healthcare environments, as well as with staffing issues and program enhancement strategies.

Security Management Consultant: Karim H. Vellani, CPP, CSC

Karim is a Certified Protection Professional (CPP) and a Certified Security Consultant (CSC). He has over 12 years of security management experience. He has extensive experience in risk assessments and security force protection for the United States Government, as well as healthcare, commercial and residential environments. Karim is currently responsible for managing the quality control/assurance function for United States Department of Homeland

Security's security protection force at over 200 federal government buildings in ten (10) states.

Technical Security Consultant: Brian Gouin, PSP, CSC

Brian Gouin is a Board Certified Physical Security Professional (PSP), a Board Certified Security Consultant (CSC), and has over 18 years of security management and security consulting experience. Prior to becoming an independent security consultant, Senior Consultant Brian Gouin, PSP, CSC owned an integration company for 12 years. For his entire 18 year career, Brian has been in charge of assessment, design, project management and/or installation of physical security devices and electronic burglary, CCTV, access control, fire and gate operator devices and systems.

Company Expertise

TAG brings several key strengths and a high level of expertise to this project. Core advantages for ABC Hospital include TAG's highly effective project management skills, our impeccable security consulting qualifications, unmatched security experience, and the independence, integrity, ethics, and confidentiality that ABC Hospital is not likely to find in other security firms.

Project Management
TAG has a track record of effective project management which results in a project that comes in on time, on budget, and with a level of quality that exceeds our clients' expectations. Our experienced security consultants operate on efficient schedules which benefit our clients in two ways: prompt response to a requests and substantial cost savings. For each project, we will develop a work plan and milestones with strict, but realistic deadlines that meet your needs.

Qualifications
Our consultants have extensive experience in private, government, and corporate situations. We have a solid reputation in the security industry of which we are very proud. Indeed, much of our business continues to come from existing clients and those they refer us to. Over the past nine years, TAG has developed a special expertise in analyzing and protecting critical assets.

Experience
TAG's consultants have a broad base of experience in security assessments, physical protection systems, and security personnel management, including significant experience in dealing with security issues involving residential communities. Our consultants have strong backgrounds not only in vulnerability assessments, but in security management, system design, program and policy

development, as well as integration of personnel, physical security, building features, and security systems (access control, intrusion detection, closed circuit television, etc.).

Independence, Integrity & Ethics
As an independent consulting firm, TAG is not affiliated with any manufacturer or vendor of security equipment, nor do we profit in any way from a client's selection of vendors or contractors. Our primary objective is to help clients provide maximum protection for their assets while obtaining the most value and benefit from their security resources. TAG's advice and recommendations are based solely on the needs of our clients. Additionally, TAG's principal consultants are members of the International Association of Professional Security Consultants (IAPSC), a renowned association known for its hallmark characteristic of independence. Throughout this project you can be assured of professional and responsive service of the highest integrity. We adhere to the Codes of Conduct and Ethics of the International Association of Professional Security Consultants (www.iapsc.org) and the American Society for Industrial Security—International (www.asisonline.org).

Confidentiality
Due to the sensitive nature of this project, all of the data collected, our analysis of that data, any recommendations to ABC Hospital resulting from our analysis, and any other information not then in the public domain received from ABC Hospital or its employees will be held in strict confidence to the extent permitted by law. Work papers and submittals to ABC Hospital will be marked "Security Sensitive Information." TAG's private and commercial contracts and projects are all subject to Confidentiality Clauses.

Company Background

Threat Analysis Group, LLC (TAG) was founded in March of 1997 as an independent security management consulting firm headquartered near Houston, Texas with offices across the country. As an independent security consulting firm, our objective is to help our clients obtain the maximum value from their security resources and expenditures. Our security assessment process is a holistic approach that examines not only the security and technology functions, but the interaction of each with policy and procedures, employee and vendor involvement, and how each element interacts with the others.

For the past nine years, TAG has provided security consulting services to hundreds of private sector companies and government agencies. TAG's consultants are Certified Protection Professionals (CPP) and Certified Security Consultants (CSC). Our security consultants are members of the American

Society for Industrial Security, International and the International Association of Professional Security Consultants (IAPSC), a renowned association known for its hallmark characteristic of independence. Throughout this project, you can be assured of professional and responsive service of the highest integrity. We adhere to the Codes of Conduct and Ethics of the International Association of Professional Security Consultants (www.iapsc.org) and the American Society for Industrial Security, International (www.asisonline.org).

References

Children's Hospital
5678 Main St.
Any Town, USA 12345
123-456-7890
Contact: Joe Jones, Director of Security

TAG is conducting a global Risk Assessment for nine (9) Children's Hospital facilities, including their main campus. The assessment includes asset identification, threat assessment and crime analysis, vulnerability assessment and security surveys, and an overall risk assessment. This project is substantially similar to the proposed ABC Hospital project.

DEF Inc.
9012 Main St.
Any Town, USA 12345
012-345-6789
Contact: Jane Jones, President

TAG has completed several security projects for DEF, Inc. throughout the United States over the past seven years. Projects have included security assessments at over 200 office buildings and archival facilities, development of security force quality assurance programs, and security policy development.

The GEF Company
3456 Main St.
Any Town, USA 12345
Contact: Ed Brown, Risk Manager
Phone: (135) 791-3579

TAG has provided security consulting services to The GEF Company, a Fortune 25 Company, for over six years. Specifically, TAG provides threat assessment and crime analysis services to GEF which assists in determining the effectiveness of GEF's risk management systems.

Project Cost

We estimate that this project will take approximately ten days to complete using a two man project team including preparation, on-site assessment activity, report development, and consultation. TAG is pleased to offer ABC Hospital a firm fixed price of $19,324 with Threat Analysis Group's standard terms and conditions. The report will be delivered within 10 days after the completion of on-site activity.

Company Contact Information

Company:	Threat Analysis Group, LLC
Point of Contact:	Karim Vellani
Address:	P.O. Box 16640, Sugar Land, TX 77496
Telephone:	(281) 494-1515
Cellular:	(713) 726-6516
Fax:	(281) 494-5700
E-mail:	kv@threatanalysis.com
Website:	www.threatanalysis.com
EIN:	01-0564411

Project Team Resumes

Karim H. Vellani, CPP, CSC

Karim H. Vellani is the President of Threat Analysis Group, LLC, an independent security consulting firm. Karim is a Board Certified Protection Professional (CPP), a Board Certified Security Consultant (CSC), and has over 12 years of security management and forensic security consulting experience.

As an independent security management consultant, Karim has been retained by Fortune 500 companies and government agencies. He has extensive experience in risk management and security force protection and provides consultation on a regular basis at government, commercial, and industrial facilities across the nation. Currently, Karim is responsible for managing the quality control/assurance function for United States Department of Homeland Security's protection force at federal government buildings in the states of Texas, Oklahoma, Mississippi, West Virginia, South Carolina, Montana, North Dakota, South Dakota, Massachusetts, and Rhode Island. Karim has also developed a Risk Assessment Methodology for Healthcare Facilities and Hospitals.

Karim provides forensic security consulting services to insurance companies and the legal profession. In this work, Karim provides litigation support to attorneys and serves as an expert witness in security related lawsuits. As an Adjunct Professor at the University of Houston—Downtown, Karim taught graduate courses in Security Management and Risk Analysis for the College of Criminal Justice's Security Management Program. He has also trained Police and Security Officers in weapons and use of deadly force and profiling and assisted in the development of maritime security training curriculum compliant with ISPS Code and the United States Coast Guard under the Maritime Transportation Security Act.

Karim is a member of the International Association of Professional Security Consultants (IAPSC) and the American Society for Industrial Security (ASIS International). He serves as an Executive Officer for the IAPSC and as the Chairman of the Professional Certification Board. Karim is also a Council Member of the ASIS International Private Security Services Council.

Education & Certifications
Master of Science, Criminal Justice Management
Sam Houston State University, August, 1998

Board Certified Security Consultant (CSC)
International Association of Professional Security Consultants (IAPSC)

Board Certified Protection Professional (CPP)
American Society for Industrial Security—International (ASIS)

Professional Affiliations
American Society for Industrial Security—International (ASIS)
International Association of Crime Analysts (IACA)
International Association of Professional Security Consultants (IAPSC)

Brian Gouin, PSP, CSC

Brian Gouin is an independent security consultant specializing in assessment, system design and project management for electronic burglary, fire, CCTV, access control, gate operator and monitoring facility systems. Brian is a Board Certified Physical Security Professional (PSP), a Board Certified Security Consultant (CSC), and has over 18 years of experience in the security and fire protection field. Brian has extensive training in system design from a vast number of manufacturers of electronic security and fire equipment.

Brian has extensive experience in electronic security hardware and software, industry installation best practices, and overall physical security. Since 2001 he has been an independent security consultant specializing in assessment, system design, and project management for electronic fire, burglary, CCTV, access control, gate operator and monitoring facility systems for all industries. Previous to that, for 12 years Brian owned a successful security installation company serving hundreds of clients which designed, sold, and installed commercial, institutional, industrial, and residential burglary, fire, CCTV, access control, and gate operator systems. The company also actively participated in the design, installation, and review of many municipal monitoring station facilities.

Brian's background in the installation field is a tremendous asset as a consultant. His experience is that of the real world and not just theory. His knowledge of the assessment, design, procurement, and construction phases of a project from all points of view helps to make the whole project go smoother and ultimately helps to achieve a better quality end product.

Mr. Gouin is a member of the International Association of Professional Security Consultants (IAPSC) and the American Society for Industrial Security (ASIS International). He is a Director of the Board for the IAPSC and serves as a Committee Member of the Professional Certification Board.

Education & Certifications
Board Certified Security Consultant (CSC)
International Association of Professional Security Consultants (IAPSC)

Board Certified Physical Security Professional (PSP)
American Society for Industrial Security—International (ASIS)

Professional Affiliations
American Society for Industrial Security—International (ASIS)
International Association of Professional Security Consultants (IAPSC)
National Fire Protection Association (NFPA)
National Association of Chiefs of Police

Document A-2 Proposal from Security Management Consulting Firm

Appendix B
Sample Service Agreement

This appendix includes a sample service agreement between the consultant and client. An attorney should be consulted before using any service agreement or contract, as laws may vary from state to state, and contract language may and should differ from consultant to consultant.

Strategic Design Services, LLC
Service Agreement

This Agreement is made as of ____________________ between Strategic Design Services, LLC (Consultant) and ____________________________ (Client). In the event of a conflict in the provisions of any attachments hereto and the provisions set forth in this Agreement, the provisions of such attachments shall govern.

1. *Services:* Consultant agrees to perform for Client the service options selected in the "Deliverables" section attached hereto and executed by both Client and Consultant. Such services are hereinafter referred to as "Services". Client agrees that Consultant shall have ready access to Client's staff and resources as necessary to perform the Consultant's services provided for in this contract.
2. *Rate of Payment for Services:* Client agrees to pay Consultant for Services in accordance with the "Project Fees" section, attached hereto and executed by both Client and Consultant.
3. *Reimbursement for expenses:* Client and Consultant agree that the "Fee" includes all applicable expenses in the performance of Services.
4. *Invoicing:* Client shall pay the amounts agreed to herein upon receipt of invoices (as outlined in the fee section) which shall be sent by, and client shall pay the amount of such invoices to Consultant. A final invoice shall be submitted upon completion of the agreed upon Service Option.
5. *Confidential Information:* Each party hereto ("Such Party") shall hold in trust for the other party hereto ("Such Other Party"), and shall not disclose to any nonparty to the Agreement any confidential information of Such Other Party. Confidential information is information which relates to Such Other Party's security, operation, trade secrets or business affairs, but does not include information which is generally known or easily ascertainable by nonparties of ordinary skill.
6. *Staff:* Consultant shall not be deemed to be an employee of the Client. Consultant is an independent contractor. Consultant assumes full responsibility for payment of any taxes due on money received hereunder. Client will not make any deductions for taxes.
7. *Use of Work Product:* Consultant and client agree that Client shall have nonexclusive ownership of the deliverable product described in the proposal and the ideas embodied therein. The consultant's notes and a file copy of all reports, documents, drawings, or other products in written, computer media or other format shall be retained in a secure manner by the Consultant.
8. *Client Representative:* The Following individual, ____________________________, shall represent the Client during the performance of this Agreement with respect to the services and deliverables as defined herein and has authority to execute written modifications of additions to this Agreement as described in Section 13.
9. *Independent Status:* The Consultant is an independent Consultant and does not represent or sell any product or service from any recommendation made herein.

Limited Warranty

10. *Liability:* Consultant warrants to client that the material, analysis, data, programs and services to be delivered or rendered hereunder will be of the kind and quality designated and will be performed by qualified personnel. Special requirements for format or standards to be followed shall be attached as an additional exhibit and executed by both Client and Consultant. Consultant makes no other warranties, whether written, oral or implied, including without limitation warranty or fitness for purpose or merchantability. In no event shall Consultant be liable for special or consequential damages, in either contract or tort, whether or not the possibility of such damages have been disclosed to Consultant in advance or could have been reasonably foreseen by Consultant and, in the event this limitation of damages is held to be unenforceable, then the parties agree that by reason of difficulty in foreseeing possible damages, all liability to Client shall be limited to One Hundred Dollars ($100.00) as liquidated damages and not as penalty.

    The Client fully understands that the Consultant is not an insurer and cannot insure the effectiveness of programs, implementation of programs, or guaranty accuracy of internal data gathered. The Client agrees that the Consultant is to gather data and recommend measures which are reasonable in nature and the lack of additional recommendations will not be construed as errors or omissions of the Consultant.

11. *Complete Agreement:* This Agreement contains the entire Agreement between the parties hereto with respect to the matters covered herein. No other Agreements, representations, warranties, or other matters, oral or written, purportedly agreed to or represented by or on behalf of the Consultant, or contained in any sales materials or brochures, shall be deemed to bind the parties hereto with respect to the subject matter hereof. Client agrees to be entering into this Agreement solely on the basis of the representations contained herein. This Agreement supersedes all prior proposals, Agreements, understandings, representations and conditions.
12. *Applicable Law:* Consultant shall comply with all applicable laws in performing Services but shall be held harmless for violation of any governmental procurement regulation to which it may be subject, but to which reference is not made in the proposal. This Agreement shall be construed in accordance with the laws of the state indicated by the Consultant's address.
13. *Scope of Agreement:* If the scope of any of the provisions in this Agreement is too broad in any respect whatsoever to permit enforcement to its full extent, then such provisions shall be enforced to the maximum extent permitted by law, and the parties hereto consent and agree that such scope may be judicially modified accordingly and that the whole of such provisions of this Agreement shall not thereby fail, but that the scope of such provisions shall be curtailed only to the extent necessary to conform to law.
14. *Additional work:* After receipt of an order that adds to the Services, Consultant may, at his discretion, take reasonable action and expend reasonable amounts of time and money based on such order. Client agrees to pay and reimburse Consultant for such action and expenditure as determined by the Consultant and agreed upon by the Client, at the time of the order.
15. *Notices:* Notices to Client should be sent to: ____________________ Notices to Consultant should be sent to: 54 South Rd. Portland, CT 06480.
16. *Assignment:* This Agreement may not be assigned any either party, without the prior written consent of the other party. Except for the prohibition on assignment contained in the preceding sentence, this Agreement shall be binding upon and inure to the benefit of the heirs, successors, and assigns of the parties hereto. Nothing in this provision prohibits the Consultant from utilizing the services of a qualified specialist-associate.

In Witness Whereof, the parties hereto have signed this Agreement as of the date written.

______________________________ ______________________________

Client **Date**

______________________________ ______________________________

Consultant **Date**

Document B-1 Sample Service Agreement

Appendix C
Sample Assessment Reports

This appendix includes two actual assessment reports that were submitted to clients with redactions. The first is from a security design consulting firm designing access control, intercom, and gate operator systems at a police department. The attachments have also been left out for confidentiality purposes. The second is from a security management firm in which the design consultant partnered with the management consultant to perform the physical security aspect of the assessment.

STRATEGIC DESIGN SERVICES, LLC

SECURITY ASSESSMENT • SYSTEM DESIGN • PROJECT MANAGEMENT

FIRE / BURGLARY / CCTV / ACCESS CONTROL / GATE OPERATORS

ASSESSMENT PHASE REPORT

SECURITY AND ACCESS CONTROL SYSTEM AT POLICE DEPARTMENT FACILITIES
CITY OF XXXXXXXXXXX

PROJECT NUMBER XXXXX

EXECUTIVE SUMMARY

In the design development phase of this project, Strategic Design Services, LLC was first asked to conduct interviews and observations at the Police Department facilities in order to design an access control, intercom and gate operator system as described in the RFP. That site visit took place on January 3rd–5th and was a success. Meetings took place with all the necessary City and Police Department personnel, different options were discussed and the Department operations were observed. All areas of both buildings and the appropriate vehicular entrances were analyzed in order to determine the necessary system coverage and design options.

From the information gathered at the site visit, industry best practices, pertinent codes and our expertise, systems were designed from a functional point of view. The details of the designs are described in this report. Products were then chosen based on the needs of the design. The products selected and the reasons for the choices are described in this report. Cost estimates for the different systems were then compiled and are included in this report along with a recommendation for breaking down the project for bidding purposes.

At the end of this report is a list of items that must be approved, agreed to or decided by the City of XXXXXXX before the construction document phase of the project can begin.

Attached to this report is the following:

- Door Schedule
- Access Control System Equipment List
- Locking Device and Door Hardware Equipment List
- Intercom System Equipment List
- Gate Operator System Equipment List
- Conceptual Drawing of Stanchion at XXXXXX Facility and Associated Pictures
- Conceptual Drawing of Gate Operator Installation at Main Facility and Associated Pictures
- Specification Sheets for Chosen Products
- Copy of Chosen Access Control Software

P.O. Box 436 • Portland, CT 06480 • (860) 342-2544 • Fax (860) 342-0542 • strategicdesignservices.com

SYSTEM DESIGN

Access Control System Design

After the interviews and observations were complete from the site visit, it was determined that 31 doors at the main facility, 6 doors at the XXXXXXX facility, the new pedestrian gate at the main facility and the rear vehicle gate at the XXXXXXX facility would have access control devices. The attached Door Schedule lists all the openings.

Each door will have a card reader on the outside of the door, except the main facility basement door to the furnace room which will have the reader on the inside of the door. The reader range for all doors will be 3″ except the main employee entrance which will have a 24″ read range. This is not the 36″ read range the PD was hoping for, but research found that 24″ was the longest range available for the application. The preferred style for the 3″ readers is "mullion" style which means the reader will be installed on the door mullion. However, if any reader requires a single gang back box, the appropriate sized reader for that application will be acceptable.

Each door will use an electric strike as the locking device. The existing strike will be used on five doors and all the rest will be new. The door hardware selection will be on a door by door basis as described in the Door Schedule. Two door hardware experts took part during the site visit to make these determinations. The basic theory is to have a door lever or pull on the outside of the door that will not operate the door unless the electric strike is released. There is then free exit from the inside of the door by operating the door lever or crash bar. Existing door hardware is being used wherever possible. Note that the crash bars and closers on the two emergency exit doors need to be serviced by XXXXXXXXXXX before any new equipment can be added.

Every door except the narcotics room door will also have a door contact installed. The purpose of this contact is to sense when a door is "propped open" for longer than a set period of time, usually a minute or so, and have an alarm go off on the software. The contacts will be recessed so they won't be seen on all but the metal doors which will have surface mounted contacts.

The reader at the main facility pedestrian gate will have a 3″ range. The locking device and hardware is dependent on the design of the gate. If the gate is designed in the same manner as the existing pedestrian gate, or the existing gate is moved over, then the strike and hardware listed in the attached equipment list is correct. If the design is different, that equipment may be different. A decision needs to be made as to that pedestrian gate design.

The reader at the rear vehicle gate at the XXXXXXX facility will have a 24″ read range to facilitate the officers driving up to the reader. The reader will be attached using modifications to the existing stanchion as detailed in the attached drawing with associated pictures. The City needs to decide whether they will make the stanchion modifications, including welding, or it will be the Contractors' responsibility. The gate operator will open when the reader is activated. The gate operator will otherwise function as it does now.

The controllers, power supplies and other equipment for the access control system will be located in the main facility ground floor 2 electrical rooms and telephone room and the XXXXXXXX communications room. An IP port and address will need to be provided in one of the electrical rooms at the main facility, the communications room at XXXXXX, and at the PC. This is a slight change from the meeting with the IT Department during the site visit, but research indicates this is the best way to design the system.

The access control system will be controlled from software on a PC, provided by XXXXXXX, in the training office. The software can allow or deny access by card, access point, time, or date. The software will also give door prop alarms and print various reports, including if access was denied and to whom. A digital camera, card printer and card printing material will also be provided for photo badge printing which is done from the same PC. The City will provide the desired camera stand.

Every authorized user will have a card with their picture and other applicable information printed on the card. They need only to hold their card in front of the reader at a correct distance and the door will release. A decision needs to be made whether the card punch slots should be horizontal or vertical and whether lanyards or clips or a combination of both should be provided for the 1000 cards.

NOTE: There is one glaring deficiency in the access control system design from a "best practices" point of view. There is no set protected area for prisoners to be brought into the facility. Right now if a prisoner got free from an officer inside the facility they could easily run out a door into the street. There should be some area with readers on both sides of the doors so that cannot happen. It is understood the Police Department is not concerned about this issue at this time and has long term plans to provide such a secure area. However, the issue is important enough to mention in this report.

Intercom System Design

There will be three internal intercom stations at the main facility: on the wall next to the communications supervisor's desk, on the wall behind communications station 2's desk, and on the wall in the Watch Commander's office. There will be one outside station at the pedestrian gate. When someone presses the button on the outside intercom station, they can be talked to by any one of the three internal stations. The three internal stations cannot talk to each other. The person on the inside can then unlock the pedestrian gate to let the person in by pressing a button or open the gate operator (covered below).

Note: If the gate operator system is not installed, the outside intercom station will move to the main employee entrance and the system will work in the same manner.

Gate Operator System Design

Per the decision made by the PD, there will be a slide gate operator at the end of the driveway going into the back parking lot of the main facility. This is depicted in the attached drawing with associated pictures. The operator itself will sit on the concrete behind the existing pedestrian gate. The gate will slide behind the existing fence, over the existing flower beds in front of the parking spaces. Four posts will be installed and a cut will be made in the existing fence on the left side of the driveway. Two posts, a section of fence and a new pedestrian gate will be installed on the right side of the driveway.

The gate will open in the morning and close in the evening at set times using a timer. There will be three safety devices for the operator: a safety loop so the gate cannot close if a vehicle is in its path, a photo beam so the gate cannot close if something else is in its path and a safety edge in case something comes in contact with the end of the gate while it is in motion.

After hours when the gate is closed, it can be opened in four ways. The first is a free exit loop so a car can drive up to the inside of the gate and it will open. The second is by push buttons in the same three places as the interior intercom stations. The third is by a wireless transmitter with a range of about 50′ that can be kept in the cars. A decision will need to be made about the number of transmitters required. The fourth is an "SOS" device where the siren from an emergency vehicle will open the gate. This is not 100% effective so it should not directly replace the wireless transmitters.

NOTE: In order for the gate operator to be installed within the project budget, the City must do a significant portion of the gate work. This includes:

1. Providing and installing the gate, gate rollers, pedestrian gate and necessary fencing.
2. Providing and installing the posts.
3. Core drilling in order to set the operator.
4. Provide power to the operator.
5. Provide conduit from the operator to building for push buttons.
6. Provide conduit from pedestrian gate to building for intercom, lock and reader.
7. Provide some sort of car stops so cars don't park in first row too close to the fence and get hit by the gate (even though there is a flower bed there).
8. Remove flowers in flower bed where gate will slide, or make other appropriate change.

Strategic Design Services, LLC will provide the City with the necessary specifications and drawings in order to provide the above Work. The Contractor will then be responsible for providing and installing all the equipment on the Gate Operator Equipment List, for running the wires within the conduit except for power, and to install the loops.

PRODUCT SELECTION

Access Control Selection

The selection of the access control system equipment was based on the system requirements and the number of installation contractors that would be allowed to bid the project. The criteria included:

1. The system had to have the capability of handling the required number of access points (doors) and to perform video badging.
2. The system had to be reasonably expandable for the future plans of the facility.
3. The system had to have user friendly software with access by point, card, time, date, etc. as well as the ability to issue temporary cards.
4. The system had to allow for a network connection to the XXXXXX facility while being programmed from the main facility.
5. The readers had to have variable read ranges to allow for the extra distance needed at the main employee entrance and XXXXXX gate.
6. The system had to fit into the budget.
7. The system had to be non-proprietary and non-geographic to allow any qualified installation company to purchase the equipment and install the system.

Based on the above criteria I found three manufacturers that met all the requirements:

1. XXXXXXXXXXXXXXXXXXXX
2. XXXXXXXXXXXXXXXXXXXX
3. XXXXXXXXXXXXXXXXXXXX

The XXXXXXX was the first system eliminated for two reasons. First, although both facilities could be programmed from the main facility, it would have to be done in a two step process instead of all at once. Second, although the system can technically accommodate up to 256 doors, after each set of 64 doors a new IP address would be needed and a third, etc. step would need to take place for programming. That is not acceptable functionality or expandability from a best practices point of view.

XXXXXXXX was chosen over XXXXXXX because SDS believes it is a better product, having seen, installed and used both. Also, XXXXXX has a much better reputation in the industry as a quality product, which actually means something in this industry. XXXXX will download all the information for both sites in one step, has a true 256 door capacity, a 65,000 user capacity, and meets all the other stated criteria. Attached with this report are specification sheets on the XXXX equipment and a copy of the XXXX software.

As for readers, none of the readers manufactured by XXXXXXXX have a longer read range than 15″. XXX and XXXXXXXX do not make their own readers. All readers and cards have to come from the same manufacturer; they cannot be mixed and matched. The longest read ranges found were between 24″ and 26″ and included the manufacturers XXXXXXXX and XXXXXX. XXXXX is the leading reader company in the industry with excellent quality, so the decision was clear to go with XXXX. The 24″ reader will be used for the two required locations and a 3″ reader for all other locations. Attached with this report are specification sheets on the XXXX readers.

Locking Device and Door Hardware Selection

There is no real functional distinction between different manufacturers of electric strikes and door hardware. The only requirement is to make sure the equipment is from a reputable manufacturer and the individual part meets the exact requirements of the door. The locking device and door hardware description in the Door Schedule and the Locking Device and Door Hardware Equipment List detail the results. Attached with this report are specification sheets on most of the hardware, though some items did not have specification sheets available.

Intercom Selection

As with the door hardware, there is no material distinction between manufacturers. The preeminent manufacturer of this kind of intercom is XXXXXXX, so that is what was selected. Attached with this report are specification sheets on the intercom equipment.

Gate Operator Selection

The first determination that needed to be made regarding the manufacturer of the slide gate operator was the style, would it be a hydraulic, chain driven or belt driven operator. Belt driven was immediately eliminated because they are of less quality and would have wear and tear issues. There is a hydraulic version at XXXXXXXX. In theory the hydraulic versions should wear less and be more reliable, however XXXXXX is having problems with the XXXXXXXX (a normally reliable brand) operator installed there. The chain driven version was chosen for the following reasons:

1. The hydraulic operator would cost about twice as much as the chain driven, and the budget is tight.
2. Because the gate will be left open all day long and it will only be opening and closing at off hours, there will not be as much wear and tear as there could be, so that will not be as much of a factor.

There is virtually no functional difference between manufacturers of slide gate operators, so the choice is a matter of personal preference. XXXXXXXXX was chosen as the manufacturer because SDS has had excellent luck with that brand both from a consultant and installation Contractor point of view. All the required safety devices and other accessories are standard for use with the XXXXXXXX operator. Attached with this report are specification sheets on the XXXXXXXX equipment and associated devices.

ESTIMATED COSTS

The estimated costs are the following:

Entire access control system	$85,500
Intercom system	$2,200
Gate operator system minus transmitters	$6,400
Transmitters	$21 each

The budget remaining for this project is about $91,600. It's hard to say exactly how competitive the pricing will be, but everyone will do their best to make sure a good number of qualified Contractors bid it to get the best pricing possible. Based on these estimates it is recommended to break the project down in the following manner for bidding purposes:

1. Main facility access control and intercom systems
2. XXXXXX facility access control system
3. Main facility gate operator system

There seems to be no reason to separate out the interior and exterior doors in the main facility access control system as was discussed. The numbers don't warrant it.

APPROVALS/DECISIONS

In order for this project to progress to the construction document phase, the following must be approved, agreed upon or decided by the City of XXXXXXXX:

1. Approve the overall design for the access control, intercom and gate operator systems.
2. Approve the equipment selection for the access control, intercom and gate operator systems.
3. Agree that the City will provide IP ports and address at two locations at the main facility and one location at the XXXXX facility as described herein.
4. Agree that the City will provide the PC for the access control software.
5. Decide whether the access cards will be punched horizontally or vertically and whether lanyards, clips or a combination of both will be used.
6. Agree that the City will provide the type of stand desired for the digital camera.
7. Approve the stanchion modification for the gate reader at XXXXXXX as described herein. Decide whether the City will make the modifications, including welding, or it will be the Contractor's responsibility.
8. Agree to service the door closers and crash bars at the two emergency exit doors at the main facility before the Contractor begins Work.
9. Agree that the City will do the construction work for the gate operator system as described herein.

10. Decide whether the new pedestrian gate design will be the same as the existing pedestrian gate, the existing gate will be moved or it will be different. If different, what will the design be?
11. Decide the number of wireless transmitters needed for the gate operator.
12. Agree on the cost breakdown for bidding purpose as described herein.

Respectfully submitted January 21, 2006,

Brian Gouin, PSP, CSC
Strategic Design Services, LLC

Document C-1 Assessment Report from Security Design Consulting Firm

Security Sensitive Information

ABC Property Owner's Association, Inc.

Security Assessment Report

July 31, 2006

Prepared by

Security Sensitive Information

Table of Contents

ABC Property Owners Association
Security Assessment Report

EXECUTIVE SUMMARY

Threat Analysis Group, LLC performed a threat assessment, conducted interviews with ABC Board Members, Safety Committee Members, and DEF Management, and performed an on-site vulnerability assessment of ABC. Based on these tasks, we identified internal and external threats to ABC, identified key assets, located vulnerable areas, and made recommendations for improving the security posture of ABC.

ABC County Police crime statistics for January 1, 2005 to June 30, 2006 indicate that robberies have increased significantly within ABC. Burglaries have risen slightly. Thefts and auto thefts have declined slightly. Assaults have neither increased nor decreased measurably. Vandalism has declined. Despite the rather unremarkable crime trends (with the notable exception of the robbery increase), ABC board members, safety committee members, and the community patrol have expressed concern about gang activity, violent crime, and property crime.

Vulnerable areas within ABC include the recreation areas, schools, and shopping centers. Of notable concern is XXXXXXXXXXXXXXXXXXXXXXXXXXXXXXX, and XXXXXXXXXXX.

Overall, we can verify most of the concerns about the Community Patrol expressed by Board Members and Safety Committee Members. In essence, the Community Patrol is ineffective, but easily corrected with strong management and supervision. While we do provide a wide range of recommendations, the bulk of the recommendations relate to improving the Community Patrol.

Summary of Recommendations

Recommendation 1: Host an ABC Crime Prevention class, with cooperation from the ABC County Police Department, wherein residents are educated on crime prevention, the role and legal limitations of the community patrol, and security awareness.

Recommendation 2: Explore mutual aid agreements with neighboring properties. Consult with an attorney before entering into an agreement.

Recommendation 3: Close the ABC building after normal business hours.

Recommendation 4: Transfer calls to the Community Patrol Officers after business hours.

Recommendation 5: Eliminate the after hours dispatcher position.

Recommendation 6: Update the Community Patrol Guidelines. The following list is non-inclusive, however the Community Patrol Guidelines and associated training should include:

- Documentation of all activities
- Patrol Procedures
- Security Incident Response Procedures
- Reporting Procedures
- Crime Response
- Trespass Enforcement
- House Check Procedures
- Emergency Action Plan Procedures
- Vulnerable Area Closings
- Conflict Resolution
- Safety Procedures
- Use of Force

— Traffic & Crowd Control
— Legal Aspects
— Report Writing
— Hazardous Materials
— Public Relations
— Effective Communications
— Fire Prevention
— Professionalism and Ethics
— Law Enforcement Liaison

Recommendation 7: Continually update Community Patrol Guidelines to reflect the latest policies, procedure, and practices.

Recommendation 8: Enforce trespassing violations.

Recommendation 9: Implement an Internal Reporting and Analysis System (electronic system preferable). This system should capture pertinent information from the Community Patrol's Daily Activity Reports (run sheets), Incident Reports, House Check Logs, and Covenant Violation Logs. The Community Services Director (CSD) should analyze this data and report it to the ABC Board. This effort will allow ABC to develop metrics to evaluate the Community Patrol. Designate an ABC Board or Committee member to oversee the reporting and analysis.

Recommendation 10: Install an Electronic "Guard" Tour System. Use a qualified and independent security consultant to design the system as described herein and to help procure and implement the system.

Recommendation 11: Install grates at the East end of the tunnel under Spring Branch Blvd and on the tunnel in Hockersmith Park.

Recommendation 12: Install "no trespassing" signs on all access points to each common area.

Recommendation 13: Install reflectors on the chains that go over the beach driveways.

Recommendation 14: ABC should explore the idea of lighting these areas with the residents who live in close proximity to the common areas.

Recommendation 15: Provide the CSD a wider range of discretion and management support to improve the Community Patrol.

Recommendation 16: The CSD should work flexible hours at least once per week to allow him/her to conduct unannounced and random night inspections of the Community Patrol.

Recommendation 17: The Community Patrol's duties should be categorized into three prioritized areas and reflected in a written mission statement:

1. Security—Pro-active crime prevention and deterrence; monitoring the dam/lake in accordance with the Emergency Action Plan; house checks; and enforcement of community guidelines and common area rules.

2. Covenants—Monitoring violations

3. Community Service—Distributing information to the Community (Board Books, etc.); Updating the community bulletin boards; Assisting resident with vehicle trouble

Recommendation 18: Consider developing a gopher/runner position to perform the Community Service aspects that the Community Patrol currently provides, such as distributing information.

Recommendation 19: Require that residents sign a Waiver of Liability when requesting house checks. Conduct house checks for residents who are out of town only, with the protection of property as the only reason for the house check. House checks should be limited to visual inspections from the patrol vehicle during the swing and night shift, with the morning shift conducting walking patrols around the houses.

Recommendation 20: Employ supervisory control of the Community Patrol to ensure that both walking and vehicular patrols are conducted in the vulnerable areas in a timely and random manner. We reiterate our recommendation to install an Electronic "Guard" Tour System to ensure these patrols are completed in accordance with Community Patrol Guidelines.

Recommendation 21: Utilize the flashing yellow lights when conducting vehicular patrols.

Recommendation 22: Change the uniform shirt color to a more visible color (white or yellow) with reflective features.

Recommendation 23: ABC Community Patrol should only patrol ABC property, and focus primarily on the common areas.

Recommendation 24: Update Community Patrol Guidelines to reflect all relevant areas discussed in this report.

Recommendation 25: Implement a system for analyzing all Community Patrol reporting and correcting deficiencies.

Recommendation 26: Maintain the Community Patrol vehicles and all equipment is in working order at all times and ensure that redundant vehicles and equipment are available.

Recommendation 27: Update Security Policies and Procedures and educate ABC residents about reporting crime to the police, not the Community Patrol.

Recommendation 28: Threat Assessment—Contact the Police Department on a monthly basis for the crime statistics for ABC and analyze this data to make necessary changes to Community Patrol duties.

Recommendation 29: Implement minimum qualifications for Community Patrol Officers:

- 21 years old
- High School Diploma
- College Preferred
- Security Experience Preferred
- State Unarmed Security Officer Registration Preferred

Recommendation 30: Require completion of the following training:

- CPR and First Aid

- State Department of Criminal Justice Services 18-hour training:
  - State Law and Regulations
  - Code of Ethics
  - General Duties and Responsibilities
  - Law
  - Security Patrol, Access Control, and Communications
  - Documentation
  - Emergency Procedures
  - Confrontation Management
- 8 hours of training and testing specific to ABC:
  - Emergency Action Plan
  - Community Patrol Guidelines
  - Covenants
  - Reporting Guidelines
- 8 hours of Field Training by a qualified Community Patrol Officer or the CSD:
  - Common Areas
  - Residential Streets
  - ABC Building

Recommendation 31: Require all Community Patrol Officers to complete additional training (continuing education) after their 12 month anniversary date and before their 18 month anniversary date.

Recommendation 32: Increase the starting wage to $10 or $11 per hour. ABC must control the wages of Community Patrol Officers.

Recommendation 33: Terminate the use of Off-Duty Police Officers.

ABC Property Owners Association
Security Assessment Report

GENERAL INFORMATION / ASSET IDENTIFICATION

ABC is a large residential community consisting of approximately 4000 homes and town-homes. The areas within ABC are primarily managed by the ABC Property Owners Association, Inc. through its agent, DEF Management Services. There are a number of common areas within ABC that are accessible for resident enjoyment including three beaches, parks, wooded areas, and golf courses. Adjacent to and within ABC are elementary schools and shopping centers. Also nearby is Forest Park High School. There are four primary vehicular access points to ABC which are public roads and accessible by through-traffic.

Safety and security concerns expressed by ABC Board Members and Safety Committee Members, and also identified during our on-site assessment relate to the ABC Community Patrol, the off-duty police officers, and lack of physical security measures. Specific concerns about each of these areas are addressed below. Also of concern is the vulnerability of common areas, an increase in crime in and around the community, and the potential for crime to negatively impact home values.

INTERVIEWS

Our initial step in the security assessment process was to discuss threats, vulnerabilities, and other key concerns with DEF Management and with ABC Board Members and Safety Committee Members. As such, the following individuals were contacted to share their thoughts on the threats and vulnerabilities at ABC and to discuss the ABC Community Patrol and Off-Duty Police Officers.

DEF Management

Interviewed:

— XXXXXX, ABC Acting Property Manager
— XXXXXX
— XXXXXX, Community Services Director

Board Members

Interviewed:

— XXXXXX
— XXXXXX
— XXXXXX
— XXXXXX
— XXXXXX
— XXXXXX
— XXXXXX

Did not request to be interviewed:

— XXXXXX
— XXXXXX
— XXXXXX
— XXXXXX
— XXXXXX

Safety Committee

Interviewed:

— XXXXXX
— XXXXXX
— XXXXXX
— XXXXXX

Not interviewed:

— XXXXXX

Sub-Association Manager
Interviewed:

— XXXXXX

SECURITY INVENTORY

Security measures generally consist of policies & procedures, physical security measures, and security personnel. ABC has two groups that serve in a security function, the Community Patrol and the ABC County Off-Duty Police Officers.

Security Policies & Procedures

Current security policies and procedures include the Community Patrol Guidelines which consist of twenty-four procedures. There is also a document entitled ABC Common Area Detail which serves as guidelines for the Off-Duty Police Officers.

Physical Security

Some common areas (Dolphin Beach) have lighting installed; however, these lights were not on during the hours of darkness. A limited number of "no trespass" signs are mounted in common areas. There are no other physical security measures currently in place for the protection of common areas within ABC.

Based on the Community Patrol 100-day Report, "DEF is currently evaluating possible use of wireless cameras at the beaches and dam that can be monitored by the dispatchers. This option is highly technical and will require several months of research and evaluation in conjunction with the appropriate committee."

Community Patrol

The ABC Community Patrol is the most visible aspect of the ABC security program. The Community Patrol is a 24-hour operation with ten full-time and part-time community patrol officers working on three shifts. Hourly rates range from $8.50 to $13.90 per hour.

Anecdotal information provided by board members, safety committee members, and community patrol officers, as well as through our on-site assessment indicates that the Community Patrol generally does not meet resident expectations. Among the more common concerns identified are a lack of liaison with the ABC County Police, not enough officers, and resident misunderstanding of the role and legal limitations of the Community Patrol.

The current responsibilities of the Community Patrol are to observe and report on activity within ABC. More specifically, they look for particular covenant violations, conduct house checks for residents, patrol the community, and deliver items to and from Board Members, Committee Members, and DEF Management. Board and Safety Committee criticisms include:

— Failure to provide good security during special events
— Spend too much time in the Lake ABC Shopping Center
— They can't get out of their Patrol vehicles
— Spend too much time acting as personal delivery people for the Board

Hiring standards and practices is another concern identified by those interviewed. Among these complaints is that the officers are too young, there is too much turnover, and a lack of background checks. Our assessment also notes that the Community Patrol Officers are not required to be licensed as unarmed security officers, nor are they required to have any security experience. Training is also a concern as there is no formal training process for Community Patrol Officers. All training is on-the-job with no formal curriculum. This has resulted in untrained Community Patrol Officers training new Officers, passing down their bad habits. Reporting is a major security function and the ABC Community Patrol is no exception. Community Patrol Officers are not given any training on effective report writing, nor are they adequately trained and supervised on their Post Orders (Community Patrol Guidelines).

Management and supervision is the major problem with the Community Patrol and has resulted in a lack of discipline, failure to carry out duties, and patrolling areas contrary to the Community Patrol Guidelines. Currently, the Community Patrol Officers report directly to the Community Services Director or the Night Supervisor. Recommendations for improving the management and supervision of the Community Patrol are found later in this report.

There are legal limitations on the Community Patrol, such as a lack of arrest authority. This has been a problem as residents have reported crimes to the Community Patrol with the expectation of a law enforcement response.

The Community Patrol is equipped with ABC Community Patrol vehicles with yellow flashing lights and police scanners, flashlights, and Nextel telephones for communications. Their uniform consists of black pants, black shoes, Maroon shirts with insignias, and optional hats. During our on-site assessment, the officers were not in proper uniform and/or their grooming was poor.

Select information from the ABC Community Patrol 100-Day Report is included in this report for the Board's consideration:

Patrolman (9)
2 × 7 days/wk   *11 pm till 7 am*
1 × 7 days/wk   *7 am till 3 pm*
2 × 7 days/wk   *3 pm till 11 pm*

Currently: 10 Patrol officers (5 full, 5 part)

Director's Position: Primary point of contact with state and local police, fire and emergency medical authorities. Equips, trains, hires and supervises the community patrol, dispatcher and beach attendant personnel and oversees the lifeguard contract. 90% of CSD's time is managing the patrol and rec attendants

Community Patrol. Operates a 24/7-patrol function of the community. Community Patrol's primary functions are:

— *Monitoring the dam/lake in-accordance-with the Emergency Action Plan—see Exhibit A.*

— *Monitoring assets of ABC to include the community building, beaches, Kids Dominion and other common areas.*

— *Deterrence of undesirable activity.*

— *First response to assist fire, medical and police. Their constant presence deters undesirable activity or if an incident occurs, they respond and notify authorities.*

— *Distributes time critical notification flyers to the community.*

- *Updates community bulletin boards notifying community of Association meetings and activities.*

Community Service. The community patrol provides as a secondary service, assistance to the community. Some examples are:

- *Barking Dog—Lost Dog—Found Dog As a safety issue we can report to animal control*
- *Vehicle out of gas—jump start vehicles—vehicle lock outs. We can assist broken down vehicles by using the patrol vehicle rotating lights and to lay down flares.*
- *Animals in house—bats—squirrels—snakes no attempt will be made to enter the house but will call animal control.*
- *Food left cooking on stove—irons left on—residents at work call officer to check.*
- *Personal items left on roofs of vehicles and resident drives off for work, briefcase, purse, c.d. players, and call and ask officers to look for property. Will be able to search the surrounding area for items.*
- *Calls to help residents with handicap person; can't get them out of vehicles or into their homes. We can offer to call fire and rescue for assistance.*
- *Resident's wheelchair is out of power—officer gets call to respond to home and plug wheelchair into outlet.*
- *Call from resident at work to check on elderly relatives.*
- *While doing walking-checks found water running out of home and notified owners.*
- *Helped residents change flat tires. Again can help by placing flares and assist with traffic.*
- *Assist residents locked out of homes. After checking identification, climbed ladders to get into upstairs windows and opened door or called for locksmith. We will be glad to offer to call a locksmith or provide a number to call.*
- *Some related assistance is provided to the Covenants function such as investigating vehicles unknown, improperly parked or appear to be abandoned; correction of minor covenant infractions such as trash, etc.*

The Community Patrol is dispatched by ABC Dispatchers located in the ABC building. Here again, we incorporate selected portions of the ABC Community Patrol 100-Day Report information on Dispatchers and the Recreation Attendants for the Board's consideration:

Dispatchers (3)

Sat-Sun	*8 am till 4 pm*
Sat-Sun	*4 pm till 12 pm*
Weekdays	*4 pm till 12 pm*

Dispatchers—after office hours services. On duty 4 PM until 12 PM Monday through Friday and 8 AM until 12 PM Saturday and Sunday. Takes calls from 703-787-SAFE and dispatches community patrol as needed or assists caller regarding questions about ABC. Monitors scheduling and use of community room—room used most evenings and weekends. Inspects community room before and after use. Configures rooms for Association meetings to include refreshments. Performs some administrative duties in support of the day staff (i.e., stuffing and mailing covenants letters, entering boat, car and recreational tags into database, coping, filing, etc.). Assists residents/visitors with:

- *Accepts request for and distributes completed resale disclosures.*
- *Makes available copies of public documents for review by the membership such as the public copy of Board packages, copies of budgets, etc. to include making requested copies.*
- *Issues boat, car and recreation tags/stickers.*
- *Assists visitors/realtors with directions, maps, questions about amenities, etc.*
- *Provides forms/applications and answers questions (i.e., PIR application, volunteer forms, etc.).*
- *Releases mail to sub-associations when called on for pickup.*
- *Provides limited administrative support to Association meetings such as making copies of documents during a Board or committee meeting.*
- *Accepts monthly assessment payments.*
- *Dispatches official documents to Directors & Officers of the Association.*
- *Accepts and processes resident complaints/requests such as maintenance issues, covenants, trash, vandalism, etc.*

During our on-site assessments, we were provided the opportunity to ride along with three Community Patrol Officers and evaluate a Dispatcher. Our findings are incorporated in the Analysis & Recommendations section below.

ABC County Off-Duty Police Officers

ABC County Off-Duty Police Officers provide ABC with some dedicated patrol of the community on Friday and Saturday from 2300 to 0300. The cost for these officers is $35 per hour. The Police Officers' responsibilities are outlined in the ABC Common Area Detail document. According to the ABC Community Patrol 100-Day Report, the *ABC County provides off duty police for dedicated patrol/services to ABC during historically high incident times, primarily weekend nights. Patrols are in uniform and marked cars. Radar is also used to deter speeding. This service has been instrumental in ticketing speeders; drunken driving arrest; response to break-ins and community disturbances at private homes as well as ABC assets such as Dolphin and West Beach; and addressing underage drinking at common area facilities.*

During our on-site assessment, we were provided the opportunity to ride along with the ABC County Off-Duty Police Officer assigned to ABC (Officer XXXXX) on July 21, 2006 from 2300 to 0012 on July 22, 2006.

THREATS

ABC Crime Statistics

Crime data for ABC obtained from the ABC County Police Department for the time period January 01, 2005 through June 30, 2006 was analyzed to determine the level and nature of crime within ABC.

The crime analysis data can be found on the CrimeAnalysis™ CD-Rom provided to XXXXXXXXX at DEF Management and via the reports provided to ABC Board President, XXXXXXXXXX.

Violent Crimes

Violent crimes, per the Federal Bureau of Investigation's Uniform Crime Reporting System, include murder, rape, robbery, and aggravated assault. During the time period analyzed, six (6) robberies were reported within the area of the ABC Security Fleet Watch. Four (4) of these incidents were reported during the first six months of 2006 and two were reported during the full twelve months of 2005. Clearly, robbery is on the rise. No murders, rapes, or aggravated assaults were reported.

Property Crimes

Property crimes, per the Federal Bureau of Investigation's Uniform Crime Reporting System, include burglary, theft, auto theft, and arson. During the time period analyzed, 59 burglaries, 189 thefts, and 11 auto thefts were reported to ABC County Police. No arsons were reported. Burglary of a Motor Vehicle (BMV) is included among the thefts.

Other Crimes

Other crimes reported from ABC during the 1.5 year period analyzed include 28 assaults, 157 acts of vandalism, five (5) unlawful carrying of weapons, three (3) sex offenses, and 10 drug abuse violations, seven (7) incidents of drunkenness, and 11 acts of disorderly conduct. The sex offenses were all acts of indecent exposure.

Discussion & Analysis

Anecdotal information provided by board members, safety committee members, and community patrol officers indicate that much of the crime is perpetrated by juvenile residents, indicating that the threat to ABC is primarily internal, not external. However, there is some anecdotal information that gang activity in the area poses an external threat. Effective crime prevention measures and deterrence should thwart much of the internal crime and our recommendations reflect a deterrence and crime prevention strategy for ABC.

Based on interviews we conducted with ABC board members, safety committee members, and the community patrol, there are four key areas of concern: gang activity, violent crime, property crime, and security/community issues. The most common crime concerns expressed by Board Members and Safety Committee Members include gang activity in the area as well as domestic assaults, burglaries, auto thefts, thefts, drug use, and vandalism in the community. Non-criminal concerns include teenagers drinking and using drugs in common areas, covenant violations, and after-hours use of common areas.

Some graffiti was seen in ABC and in the 7/11 shopping center, though none of it appears to be gang related. While some of the concerns expressed above are verified in the crime statistics provided by the ABC County Police, many of the concerns are not crime related. The community issues cannot be resolved by law enforcement, but rather only through agents of ABC. In other words, a proprietary (community patrol) or contract security force is a viable solution. Both of these alternatives are discussed in the Analysis & Recommendations section below.

Officer XXXXXXX, an ABC County off-duty police officer who works the ABC detail, informed us that there is not a lot of crime in ABC and that the other police officers assigned liked this detail because all they have to do is drive around. Officer XXXXXXX laughed at the notion that there is a lot of gang and drug activity in ABC, stating specifically that it was not true. While Officer XXXXXXX's comments are anecdotal, we do take them into consideration as we did all threat related comments.

ABC Property Owners Association
Security Assessment Report

VULNERABLE AREAS/ASSETS
Vulnerable areas within ABC include the recreation areas, schools, and shopping centers. Throughout this report, we generally refer to these areas as the Common Areas. Recreation areas include Hockersmith Park, Ann Mocure Wall Park, the Country Club, Southlake's Recreation Center, Kid's Dominion Park, West Beach, Dolphin Beach, Beaver Landing, wooded areas, golf courses, and trails. Key areas of concern identified during our on-site assessment include:

— XXXXXX—largest beach with high traffic and most community amenities
— XXXXXX—boat storage
— XXXXXX—Gazebo, BBQ Pits, Picnic Tables
— XXXXXX—hidden from view
— XXXXXX—partial obstruction of view and area for school children to congregate
— XXXXXX—large opening hidden from view

Generally, all of these areas are dark during the nighttime and are not currently patrolled adequately by the Community Patrol or Off-Duty Police Officers. It should also be noted that some of these areas are not ABC properties, nor are they patrolled by the ABC Community Patrol.

ANALYSIS & RECOMMENDATIONS
General
Overall, we find that a more visible security presence is needed within ABC and this concept is reflected throughout many of our recommendations. Existing protections systems are currently not effective despite a solid foundation and funding for effective crime prevention. One key issue identified during our interviews needs to be addressed: There appears a general misunderstanding among residents regarding the role and legal limitations of the ABC Community Patrol.

Recommendation 1: Host an ABC Crime Prevention class, with cooperation from the ABC County Police Department, wherein residents are educated on crime prevention, the role and legal limitations of the community patrol, and security awareness.

Working with adjacent properties can also help increase security within ABC; however, this should be a carefully orchestrated endeavor and ABC must conduct its due diligence and consult with legal counsel. The ABC Community Patrol should not patrol non-ABC properties until the liability exposure has been considered and an attorney consulted.

Recommendation 2: Explore mutual aid agreements with neighboring properties. Consult with an attorney before entering into an agreement.

The ABC Building is open after business hours with a dispatcher, typically female employees, available until 2200. Based on our observation of the dispatcher function, the low level of dispatched calls received by the Community Patrol, the low level of traffic after business hours in the building, and the limited need for the office to be open after hours, we find that keeping the ABC building open after business hours is an unnecessary vulnerability.

Recommendation 3: Close the ABC building after normal business hours.

Recommendation 4: Transfer calls to the Community Patrol Officers after business hours.

Recommendation 5: Eliminate the after hours dispatcher position.

Policies & Procedures
The Community Patrol Guidelines need to be updated to reflect the expectations of the ABC Board and residents, and the recommendations outlined in this report.

Recommendation 6: Update the Community Patrol Guidelines. The following list is non-inclusive, however the Community Patrol Guidelines and associated training should include:

- Documentation of all activities
- Patrol Procedures
- Security Incident Response Procedures
- Reporting Procedures
- Crime Response
- Trespass Enforcement
- House Check Procedures
- Emergency Action Plan Procedures
- Vulnerable Area Closings
- Conflict Resolution
- Safety Procedures
- Use of Force
- Traffic & Crowd Control
- Legal Aspects
- Report Writing
- Hazardous Materials
- Public Relations
- Effective Communications
- Fire Prevention
- Professionalism and Ethics
- Law Enforcement Liaison

Recommendation 7: Continually update Community Patrol Guidelines to reflect the latest policies, procedure, and practices.

Recommendation 8: Enforce trespassing violations.

Recommendation 9: Implement an Internal Reporting and Analysis System (electronic system preferable). This system should capture pertinent information from the Community Patrol's Daily Activity Reports (run sheets), Incident Reports, House Check Logs, and Covenant Violation Logs. The Community Services Director (CSD) should analyze this data and report it to the ABC Board. This effort will allow ABC to develop metrics to evaluate the Community Patrol. Designate an ABC Board or Committee member to oversee the reporting and analysis.

Physical Security
During our on-site assessment, the need for physical security measures was evaluated, both electronic and non-electronic. The comments from those interviewed and our own expertise and industry best practices were taken into consideration in making the following recommendations.

Electronic "Guard" Tour System
There are areas within the property that the community patrol needs to patrol on foot on a regular basis, including the vulnerable areas identified earlier in this report. Vehicular patrols are not effective for most of these vulnerable areas. As such, we recommend walking patrols of the common areas and a system for verifying the patrols are occurring as specified. This will enable ABC to not only deter and prevent crime, but also to evaluate the Community Patrol's performance. The most effective way to accomplish this goal is with the installation of an Electronic "Guard" Tour System.

ABC Property Owners Association
Security Assessment Report

Operation of the Electronic "Guard" Tour System is as follows: Small chips are installed at various points throughout the property. These chips do not require any power. Special vandal resistant enclosures need to be designed to protect the chips. Each Community Patrol Officer has a handheld "wand" which they carry and hold up to the chip. The wand records the chip location and time. At the end of each shift, the Community Patrol Officer downloads the information on the wand via a caddy connected to a PC. The information can then be collected and analyzed by an authorized user on the PC. The following locations within ABC property are recommended as a chip location:

1. At the High School end of the path to the High School at the end of Olivia Way.
2. After the trees on the path to the High School at the rear of the Southlake recreation parking lot.
3. Pool house at the Southlake Recreation Center.
4. West Beach down the end of the path where the boat racks are.
5. Pavilion at Dolphin Beach.
6. Dam pump house at Dolphin Beach.
7. End of the path on the far side of the dam at the end of Golf Club Rd.
8. In the rear portion of Hockersmith Park.
9. End of Northgate St.
10. Gazebo at Beaver Landing.
11. ABC Administration Building.
12. Up the path at Kids Dominion Playground.
13. Down by the stream at the East end of the tunnel under Spring Branch Blvd.
14. Down by the water at the Timber Ridge turnaround.

If ABC is able to obtain mutual-aid agreements with adjacent properties and determines that Community Patrol Officers are to patrol off ABC property, the following locations should have a chip installed after permission is granted by the owner of the other property. As discussed later, we do not specifically recommend patrolling non-ABC properties; however, there may be valid reasons for reaching an agreement with neighboring properties.

1. Burger King Plaza
2. Pattie Elementary School
3. Henderson Elementary School
4. ABC Elementary School
5. 7-11 Shopping Center
6. Building at the tennis courts
7. Country Club
8. Ann Moncure Wall Park

The estimated cost to install such a system is $8000. Design and project management costs are not included in this cost.

Recommendation 10: Install an Electronic "Guard" Tour System. Use a qualified and independent security consultant to design the system as described herein and to help procure and implement the system.

Grates, Signage, and Reflectors

There are security issues with unwanted juvenile activity or in some cases criminal activity in the tunnel under Spring Branch Blvd. and the tunnel in Hockersmith Park. Currently there is a grate covering the West end of the Spring Branch tunnel that would keep individuals out.

Currently there are signs at the beaches and parks that say they are closed at dusk. However, there are limited "no trespassing" signs located in these common areas. This makes it difficult for law enforcement to enforce trespassing violations. After dark, it is very difficult to see the chains which block vehicular entry to the beach driveways and is a safety hazard.

Recommendation 11: Install grates at the East end of the tunnel under Spring Branch Blvd and on the tunnel in Hockersmith Park.

Recommendation 12: Install "no trespassing" signs on all access points to each common area.

Recommendation 13: Install reflectors on the chains that go over the beach driveways.

Lighting
There is no lighting in most common areas except Dolphin Beach, which are not turned on at night. There are advantages and disadvantages to installing lighting in these areas:

Pros

- From a strict security point of view, lighting is critical and is a good deterrent to crime.
- The community patrol would be better able to patrol these areas.

Cons

- Lighting at these areas can create a nuisance for the residents in close proximity to the common areas. Low post lights or ground lights may be advantageous, but will reduce lighting effectiveness.
- The high cost of installing lights in these areas.

Recommendation 14: ABC should explore the idea of lighting these areas with the residents who live in close proximity to the common areas.

Closed Circuit Television System (Cameras)
We were specifically asked to determine if a closed circuit television system (CCTV) would be effective in ABC, especially in the common areas. A CCTV system by itself is not a preventive measure, as it will deter only the most easily deterrable criminals. To serve as a preventive measure, the CCTV system must be monitored and suspicious events responded to quickly. Beyond this security principle, a CCTV system is prohibitive for five reasons:

1. Light—Cameras require some light. While "no light" cameras exist, they are expensive and installing the necessary lights would be a more cost effective alternative.

2. Power—Cameras need power to operate. There is no power at two of the Beaches or the parks. There is power at the pump house at Dolphin Beach, but it is a good distance from any potential camera location.

3. Signal Transmission—The signal needs to be transmitted from the cameras to a location for recording. This can be accomplished by direct wire, network or wireless (line of sight). ABC has no buildings at any of these facilities to install recording equipment, which means the signals need to be sent to a central location. ABC currently has no means at the locations for transmission.

4. Cost—The cost to install cameras at these locations would not be cost effective.

5. Monitoring—In order for the CCTV system to be a preventive measure, it must be monitored with a security response to suspicious activity. There is no mechanism in place to monitor the system, nor do we recommend it.

We do not recommend the installation a CCTV system at this time for the reasons cited above.

Community Patrol

General

We recommend continuing the ABC Community Patrol service, but take the necessary steps to increase its effectiveness, both in their customer service role and their security role. We find the number of Community Patrol Officers per shift to be adequate. The primary issue with the community patrol today is a severe lack of management and supervision. The Community Services Director (CSD) is ultimately responsible for the Community Patrol.

Recommendation 15: Provide the CSD a wide range of discretion and management support to improve the Community Patrol.

Recommendation 16: The CSD should work flexible hours at least once per week to allow him/her to conduct unannounced and random night inspections of the Community Patrol.

Recommendation 17: The Community Patrol's duties should be categorized into three prioritized areas and reflected in a written mission statement:

1. Security—Pro-active crime prevention and deterrence; monitoring the dam/lake in accordance with the Emergency Action Plan; house checks; and enforcement of community guidelines and common area rules.

2. Covenants—Monitoring violations

3. Community Service—Distributing information to the Community (Board Books, etc.); Updating the community bulletin boards; Assisting resident with vehicle trouble

Recommendation 18: Consider developing a gopher/runner position to perform the Community Service aspects that the Community Patrol currently provides, such as distributing information.

A potential problem was identified during our on-site assessment wherein the Community Patrol was conducting a house check for a resident who was in town, but under threat of violence. In this instance, the house check was requested to keep an eye out for the resident's estranged husband who had apparently threatened her with violence. This exposes ABC to liability for accepting this duty should anything happen to this resident.

Recommendation 19: Require that residents sign a Waiver of Liability when requesting house checks. Conduct house checks for residents who are out of town only, with the protection of property as the only reason for the house check. House checks should be limited to visual inspections from the patrol vehicle during the swing and night shift, with the morning shift conducting walking patrols around the houses.

Responsibilities

Currently, the Community Patrol is spending little time in the security function, while much of its time is spent on the community services function. During our on-site assessment and based on discussions with Community Patrol Officers, we learned that the Community Patrol Officers rarely conduct walking patrols and when they do, they do so in pairs for safety reasons. We identified no direct threat to their safety that would prohibit them from conducting walking patrols.

In order to increase the effectiveness of the Community Patrol and given their limited legal authority, the Community Patrol's primary mission must be to be highly visible in order to deter and prevent crime. This includes highly visible, random vehicular patrols; highly visible, random walking patrols; and a professional, authoritative appearance.

Recommendation 20: Employ supervisory control of the Community Patrol to ensure that both walking and vehicular patrols are conducted in the vulnerable areas in a timely and random manner. We reiterate our recommendation to install an Electronic "Guard" Tour System to ensure these patrols are completed in accordance with Community Patrol Guidelines.

Recommendation 21: Utilize the flashing yellow lights when conducting vehicular patrols.

Recommendation 22: Change the uniform shirt color to a more visible color (white or yellow) with reflective features.

Recommendation 23: ABC Community Patrol should only patrol ABC property, and focus primarily on the common areas.

Recommendation 24: Update Community Patrol Guidelines to reflect all relevant areas discussed in this report.

Recommendation 25: Implement a system for analyzing all Community Patrol reporting and correcting deficiencies.

Equipment

The equipment available to the ABC Community Patrol is sufficient when in proper working order. Currently, the Community Patrol uses marked patrol vehicles with flashing yellow lights, Nextel phones for communications, and flashlights. The Community Patrol uses several forms while carrying out their duties including house check logs, run sheets (daily activity reports), and incident reports. While these reporting mechanisms are in place, there is generally poor reporting and no analysis of the information generated by the Community Patrol.

Recommendation 26: Maintain the Community Patrol vehicles and all equipment is in working order at all times and ensure that redundant vehicles and equipment are available.

Crime Response & Law Enforcement Liaison

The Community Patrol should not respond to crime. Their function is to prevent and deter crime. Once a crime has occurred, its must be reported to the ABC County Police Department. The Community Patrol, via the CSD, should maintain liaison with the ABC County Police Department. This liaison should include monthly crime statistics review as well as area crime information.

Recommendation 27: Update Security Policies and Procedures and educate ABC residents about reporting crime to the police, not the Community Patrol.

Recommendation 28: Threat Assessment—Contact the Police Department on a monthly basis for the crime statistics for ABC and analyze this data to make necessary changes to Community Patrol duties.

ABC Property Owners Association
Security Assessment Report

Selection & Training
Turnover and the overall effectiveness of the Community Patrol are highly dependent on the quality of personnel hired, the level of training provided, and the wage level. Currently, there are no hiring standards nor any formal training provided to Community Patrol Officers. Community Patrol Officer wages range from $8.50 per hour to $13.90 per hour and are dictated by DEF, not ABC. The lower end of this wage range does not allow ABC to attract highly qualified applicants, especially in XXXXXXXXX's tight security labor pool.

Recommendation 29: Implement minimum qualifications for Community Patrol Officers:

- 21 years old
- High School Diploma
- College Preferred
- Security Experience Preferred
- XXXXXXX Unarmed Security Officer Registration Preferred

Recommendation 30: Require completion of the following training:

- CPR and First Aid
- XXXXXX Department of Criminal Justice Services 18-hour training:
  - XXXXXXX Law and Regulations
  - Code of Ethics
  - General Duties and Responsibilities
  - Law
  - Security Patrol, Access Control, and Communications
  - Documentation
  - Emergency Procedures
  - Confrontation Management
- 8 hours of training and testing specific to ABC:
  - Emergency Action Plan
  - Community Patrol Guidelines
  - Covenants
  - Reporting Guidelines
- 8 hours of Field Training by a qualified Community Patrol Officer or the CSD:
  - Common Areas
  - Residential Streets
  - ABC Building

Recommendation 31: Require all Community Patrol Officers to complete additional training (continuing education) after their 12 month anniversary date and before their 18 month anniversary date.

Recommendation 32: Increase the starting wage to $10 or $11 per hour. ABC must control the wages of Community Patrol Officers.

ABC County Off-Duty Police Officers
We found several problems with the Off-Duty Police Officers assigned to ABC. First, there are no metrics by which to judge their value to ABC. They have been asked repeatedly by DEF Management to provide reports of their activities while patrolling ABC and none have been provided. Second, based on our ride-along with the police officer on July 21, 2006, it was determined

Security Sensitive Information 19
Page

that the police officer is only providing limited police services (traffic enforcement) and does not serve as an effective deterrence to crime as he failed to patrol ABC's vulnerable areas. Third, the police officers are not cost effective, costing ABC approximately $15,000 annually. These funds could be redirected to enhance the Community Patrol. Fourth, the police officers have left ABC on occasion to transport a prisoner or to respond to a call outside ABC. ABC County has expected full payment despite the lack of full service during the hours scheduled. Anecdotally, one police officer requested an additional two hours of pay to process an arrest made in ABC that took him until 0500. Additionally, one of the night shift Community Patrol Officers informed us that the off-duty police officer had left ABC once a weekend for the last three weekends for approximately two hours. Fifth, if the Community Patrol recommendations are implemented, the community patrol would provide a significantly more cost-effective solution than the Off-Duty Police.

Recommendation 33: Terminate the use of Off-Duty Police Officers.

We recognize that this recommendation may cause some degree of animosity by the ABC County Police Department; however, the relationship between ABC and *ABC* can be reestablished over time. To speed up that process, ABC County Police could be hired for the special events protection, namely those where a security contractor (XXXXXXX) has provided service.

If ABC determines that the Off-Duty Police Patrol should continue despite our recommendation, we recommend that the Off-Duty Police Officers are provided ABC-specific training, required to follow the ABC Community Patrol Guidelines, and provide all necessary reporting for metrics development.

Alternatives to Community Patrol
We do not recommend an armed security contractor due to a lack of historical threats and the liability exposure that would result from an armed security force. The community patrol could be replaced with an unarmed contract security force; however, we do not recommend this for several reasons. ABC, with 280 man-hours per week of coverage, would be a relatively small contract for a quality security contractor, and thus not a priority client. This presents two problems. First, the security personnel assigned to ABC would likely be newer employees with limited experience. In essence, ABC would be the proving grounds for the security officer. Second, quality security officers would be reassigned to the contractor's more high profile clients causing turnover at ABC. While it is possible to overcome these security contractor problems if ABC paid higher rates, it would not be cost-effective. If the community patrol recommendations are implemented, the Community Patrol would provide a significantly more cost-effective solution than a security contractor.

The use of a security contractor during special events and in other limited circumstances may be warranted to supplement the ABC Community Patrol.

Document C-2 Assessment Report from Security Management Consulting Firm

Appendix D
Sample Design Specifications

This appendix shows the actual design specifications that were used for an IFB along with the design drawings in Appendix E. A well-respected integrator called these exact specifications "the best specifications he had ever read," one of the more rewarding career moments for this author.

TECHNICAL SPECIFICATIONS

PART I GENERAL

1.1 DESCRIPTION

A. This specification calls for the furnishing of labor, shop engineering, supervision, materials, methods, connection, and testing and of access control, burglary and CCTV systems at the ABC Company facility at 1234 Main St., Anywhere, USA 12345. The systems shall include, but not be limited to, controllers, readers, electric strikes, REX devices, system software, door contacts, door releases button, motion detectors, keypads, panic buttons, glass break detectors, temperature sensors, cameras, digital recorders, monitors and power supplies.

B. Contractor shall furnish all equipment, material, transportation, permits, permit fees, and labor necessary to provide complete system Work, in first class condition, as indicated on the Drawings and specified herein, including all devices and wiring.

C. The Owner shall be responsible for 120 VAC terminations for all necessary controllers and power supplies as well as providing receptacles for the recorders, monitors, burglar alarm panel, and HUB. The Owner shall provide the central processing unit (CPU) and the access cards for the system. Contractor shall program the cards into the system. The Owner shall stub up conduit for each card reader and provide the voltage monitor, electric lockset and necessary electric hinges and door boring work in order for the Contractor to install the electric strikes in the double doors.

1.2 DEFINITIONS

A. Contractor: This term designates the company which conducts the Work and is responsible to insure the Work is provided. This term specifically refers to a company that is qualified to perform the Work specified herein relating to the installation of the system.

B. Consultant: This signifies the design consulting firm representing the Owner in matters relating to the design and specifications of the systems and responsible for the review and approval of Contractor's submittals, approval of final testing and similar and other related matters and duties. The Design Consultant is Strategic Design Services, LLC.

C. Owner: The Owner is ABC Company. The use of the term "Owner" is intended to signify that entity that has the authority to approve. Owner's authorized representative(s) shall be entitled to any and all rights assigned herein and elsewhere to Owner.

D. Work: This term denotes the entire completed construction required to be furnished herein. Work is the result of performing services, furnishing labor, and furnishing and incorporating material and equipment into the construction.

1.3 OVERVIEW

A. An access control system shall be installed to include proximity card readers, controllers, electric strikes, flush mount door contacts for "door ajar" monitoring, REX motions detectors, a door release button and a Hub as described herein.

B. The access control software shall be installed and programmed on the Owner supplied PC.

C. A burglar alarm system shall be installed to include a panel, keypad, glass break detectors, dual technology motion detector, temperature sensors, and a panic button. The burglar alarm system and access control system shall be integrated together as described herein.

D. Central station monitoring services shall be provided for the burglar alarm system.

E. A CCTV system shall be installed including fixed dome cameras, digital recorders, monitors, and power supplies as described herein.

F. The software for the digital recorder shall be given to the Owner but not installed on a PC as part of this project.

1.4 RELATED DOCUMENTS

A. Each and every Contractor and/or supplier providing goods or services referenced in or relating to these specifications shall also be bound by these specifications and attached Drawings.

B. Attached Drawings listed below apply to and are a part of these specifications:

Name	Drawing Number(s)
Cover Sheet	SE-0
Security Layout	SE-1
Door Schedule/Door Details	SE-2
Access Control/Burglary Riser Diagram	SE-3
Access Control Details	SE-4
Camera System Riser Diagram	SE-6

1.5 INTENT

A. It is the intent of the Specifications and Drawings to call for finished Work, tested, ready for operation and programmed for the specifics of this project and Owner's requirements.

B. The Contractor shall be responsible for providing any incidental accessories necessary for complete Work even if not particularly specified without additional cost. No exclusion from, or limitations in, the language used in the drawings or specifications shall be interpreted as meaning that the appurtenance or accessories necessary to complete any required system or item of equipment are to be omitted.

C. Minor details not usually shown or specified but necessary for complete installation and operation shall be provided without additional cost.

D. In all cases wherein apparatus herein referred to in singular numbers, it is intended that such reference include as many such items as are required to complete the Work.

E. Work under the jurisdiction of the State and local Fire Marshal as it relates to electrically locked door control systems shall comply with requirements set forth by the Fire Marshal and local and State building, fire and electrical codes.

F. In brief, the scope of the Contractor's Work is as follows:

1. Provide all requested information and qualifications with the bid.
2. Provide submittals and shop drawings as specified.
3. Provide all equipment, cable, conduit and incidentals specified or required for each specific system.
4. Install all system equipment, conduit, cable, etc. for each specified system in accordance with the project construction schedule.
5. Provide competent project management, attend job conferences and construct continuation drawings as required.
6. Provide weekly progress reports to the Owner.
7. Program all systems and load data bases as directed by the Specifications and custom requirements as developed by the Owner in conjunction with the Consultant.
8. Fully test all systems and provide testing documentation including testing in presence of the Consultant.
9. Provide Owner training.
10. Provide customized operations and maintenance manuals and "as built" drawings.
11. Provide a Central Station Monitoring contract for the burglar system.
12. Provide first year maintenance and service to maintain first year guaranty and warranty as part of lump sum price.

G. Nothing contained in the drawings and specifications shall be construed to conflict with applicable State and local laws, codes, and ordinances. Comply with drawing and specification requirements which are in excess of minimum code requirements.

H. The drawings of necessity utilize symbols and schematic diagrams to indicate various items of Work. None of these have any dimensional significance nor do they delineate every item required for the intended installations. The Work shall be installed in accordance with the intent diagrammatically expressed on the drawings and in conformity with the dimensions expressed on equipment shop drawings. No interpretation shall be made from the limitations of symbols and diagrams that any elements necessary for complete Work are excluded.

I. Certain details appear on the drawings which are specific with regard to the dimensioning and positioning of the Work. These details are intended only for the purpose of establishing general feasibility. They do not obviate field coordination for the indicated Work.

1.6 CONTRACTOR QUALIFICATIONS

A. The Contractor shall be a bona fide and technically experienced contractor, licensed by the State of XXXXXX for the installation of low voltage and signal communications systems, and engaged in security system maintenance, service and contracting for at least the last 5 years.

B. The Contractor shall have a strong Info Graphics background with GE Security Sapphire Pro experience.

C. The Contractor shall be a union contractor.

D. Any Contractor who intends to submit a bid must include the following qualification data in writing. The requirements are in addition to those requirements in the entire specifications. Grounds for disqualification shall exist if it is believed that the information submitted is inaccurate or, in the opinion of the Owner or Consultant, does not satisfy the requirements.

1. Proof that the Contractor has a minimum of five years' experience in successfully completing projects of similar scope and equipment applications.
2. Proof that the Contractor is a firm which currently and regularly installs, services, and maintains security systems as a full time business.
3. Provide a list of five similar projects that have been completed by the Contractor and are operational for a minimum of one year. For each facility, list name, description, dollar value, location of the installation, date of contract completion, and contact name and telephone number. At least three of these projects must include Sapphire Pro or Diamond II software.

4. Proof that the Contractor and any Sub-Contractors are licensed by the State of XXXXXX for the Work being provided.
5. Proof of insurance for the Contractor and any Sub-Contractors.
6. If the firm has been involved in litigation or criminal action with a client, City, County, State, or Federal government agency within the past five years, provide full details and status of each occurrence.

1.7 SUBMITTALS

A. Prior to assembling or installing the Work, the Contractor shall prepare and submit within 15 days of contract award two copies of shop drawings to the Consultant for review and approval in accordance with the general conditions and as herein specified. The shop drawing shall be clearly presented and include sufficient information to determine compliance with drawings and specifications.

   1. Shop drawings shall include complete installation drawings including system block and functional diagrams of all subsystems and integrated systems and terminal point to point wiring diagrams for each type of device including correct terminal or connector pin designations. Wiring diagrams shall include all components of the system and each component is to be labeled on the diagrams.
   2. Shop drawings shall include large scale plans illustrating the layout of equipment in equipment rooms.
   3. Shop drawings shall include wiring and conduit layouts.
   4. Submit catalogue information, factory assembly drawings, and field installation drawings as required for complete explanation and description of all items of equipment.
   5. Include a detailed description of hardware to be supplied and a description relating hardware to functional diagram.

B. At the conclusion of the project, the Contractor shall provide "as built" documentation to consist of two sets of drawings. The "as built" drawings shall be a continuation of the Contractor's shop drawings as modified, augmented, and reviewed during the installation, check out and acceptance phases of the project.

1.8 ARRANGEMENT OF WORK

The drawings are partially diagrammatic and indicate general arrangement of the Work. The drawings may not show the exact locations of devices to be installed. Prior to assembling or installing the Work, the Owner must approve all device locations.

1.9 PROJECT CONFERENCES

A. Prior to the start of installation, the Contractor and Owner will attend a job coordination meeting. The intent of this meeting is to insure complete coordination of the Work.

B. The Contractor shall be responsible for attending job conferences as required by the Owner or Consultant. The Contractor shall provide a fully qualified Project Manager to attend said meetings. This representative shall be fully authorized to represent the Contractor in all contract discussions.

C. The Contractor shall provide weekly project progress reports in writing to the Owner and Consultant.

1.10 SUBLETTING

A. There shall be a single Contractor responsible for the proper installation and testing of all systems specified.

B. The Contractor must submit the names of any sub-contractors to the Consultant for approval before letting any sub-contractors perform Work. In no case shall such consent be construed as a release from the obligations of the Contractor.

C. The Contractor shall be responsible for familiarizing each sub-contractor with the aspects of the Work and shall be responsible for coordinating the Work of the sub-contractor to prevent any interference or omission whatsoever.

1.11 VERIFYING JOB-SITE CONDITIONS

It shall be the responsibility of the Contractor to examine job-site conditions prior to submitting a bid and to become thoroughly familiar with actual existing conditions at the sites. The intent of the Work is shown on the drawings and described herein, and no consideration will be granted by reason of lack of familiarity on the part of the Contractor with actual physical conditions at the sites.

1.12 CHANGES IN SCOPE/WORK REQUIREMENT

The Contractor is advised that no changes in the scope of work shall be made without written authorization from the Owner and Consultant and formal modification of the specifications.

1.13 PRODUCT HANDLINGS

A. Equipment and materials shall be properly stored, adequately protected and carefully handled to prevent damage before and during installation. Equipment and materials shall be handled, stored and protected in accordance with the manufacturers' recommendations. Equipment provided with a factory finish shall be maintained free of dust, dirt, and foreign matter. Dents, marred finishes, and other damage shall be repaired to its original condition or shall be replaced at no additional cost to the Owner.

B. Any material, device, or equipment damages prior to or during installation and before acceptance of the completed system by the Consultant shall be replaced unless repairs can be made that are acceptable to the Owner. Any such replacement or repairs, including repairs to the finish, shall be made at no cost to the Owner.

1.14 SECURITY

The Contractor shall be solely responsible to provide protection of his equipment, materials, and tools. Contractor shall bear the full responsibility of any theft or mysterious losses or damage to his equipment, materials, or tools. Insurance coverage in these regards, if any, as well as the cost of such coverage, shall be the responsibility of the Contractor.

1.15 TEMPORARY SECURITY/PROTECTION PROVISIONS

A. The Contractor is required to provide, during installation of the Work, temporary security and protection provisions including, but not limited to, barricades, cones, warning signs and similar provisions. They are intended to minimize personal injuries, property losses, and claims for damages at working site.

B. The Owner assumes no responsibility for temporary security/protection provisions.

1.16 GUARANTEE

A. All Work performed and all material and equipment furnished under this contract shall be free from defects in materials and/or workmanship and shall remain so for a period of at least one (1) year from the date of acceptance. Upon receipt of notice from the Owner of failure of any part of the guaranteed system, the Contractor shall promptly restore the defective component to provide an acceptable system at no cost to the Owner.

B. The guarantee period shall commence upon acceptance by the Owner.

C. The full cost of maintenance, labor and materials required to correct any defect during this one-year period shall be included in the submittal bid. During the guarantee period, there shall be no charges to the Owner for service calls (mileage, labor, travel, expenses, etc.) for guarantee work.

1.17 DEFECTS

A. Should any material furnished and installed fail to comply with the specifications, it will be replaced at Contractor expense.

B. Only new products and equipment shall be installed. Refurbished, reconditioned, seconds, repaired or components used in any way shall be unacceptable.

1.18 REFERENCE STANDARDS

The publication, codes and standards listed below form a part of this specification. They shall be adhered to and shall not be contravened without the approval from the Consultant or Authority Having Jurisdiction (AHJ). All standards and publications shall be the most current version.

1. All Federal, State and Local codes governing (latest issue)
2. Americans with Disabilities Act (ADA)
3. American National Standards Institute (ANSI) publications
4. NFPA 70 National Electrical Code
5. NFPA 101 Life Safety Code (especially sections on means of egress)
6. NFPA 72 National Fire Alarm Code
7. National Electrical Manufacturer's Association (NEMA) publications
8. American Society for Testing & Materials (ASTM)
9. Underwriters Laboratories, Inc. Standard for Safety UL 294 Access Control System Units
10. FCC—Class B Part 15

1.19 EQUIPMENT SUBSTITUTIONS

There shall be no equipment substitutions allowed for equipment specified.

1.20 OWNER TRAINING

A. The Contractor shall be responsible for training a minimum of 3 people designated by the Owner on the operation and maintenance of all equipment installed.

B. Training shall include instruction in the manual and automatic operation of all system components and equipment. Training shall be provided by a technician familiar with all software, subsystems and integrated system operation for a period of no less than eight (8) hours of instruction.

C. Scheduling of training shall be prior to completion of the project and coordinated with the Owner a minimum of 10 days prior to commencing same. The Contractor shall provide a training agenda to the Owner for review prior to any scheduling.

PART II PRODUCTS

2.1 SYSTEM COMPONENTS

A. All materials shall be strictly in accordance with the manufacturer's model numbers, performance levels, quality, style and sizes as specified herein. Manufacturers' names and model numbers are given in the Specifications for the purposes of establishing a standard of performance, quality, style, size and type. No substitutions are allowed.

B. When a specified manufacturer's product has been superseded by a newer model, the later model shall be furnished, provided the newer model retains the essential characteristics of the item specified herein and maintains compatibility with integrated systems hardware and software.

C. The materials and equipment to be furnished shall be new and unused and fabricated from new materials. Factory "reconditioned" products are not allowed.

D. Materials and equipment shall be the standard product of a manufacturer regularly engaged in the production of the required type of material or equipment for at least five (5) years and shall be the manufacturer's latest design with published properties.

2.2 ACCESS CONTROL SYSTEM SPECIFICATIONS

A. Access Control System Operational Features

1. Card Access Control:
    a. The security access control system shall provide the following card access control operational objectives:
        1. Controlled entry, via access card readers, of only authorized personnel to secured areas based on cardholder information entered and stored in the system database.
        2. The access request response time from card presentation, database verification, to electric lock/unlock shall be no more than one second in normal operating mode on a fully loaded system.
        3. All access requests, both authorized and denied, shall be sent to the host for storage and annunciation, as required, with the cardholder number, name, and access point/area where access was attempted or gained.
        4. The software package shall provide for global and local anti-passback, and also provide a facility for

soft anti-passback (i.e., allowing entry following an anti-passback violation but still report and log the violation). The system shall also be capable of providing timed anti-passback at individual readers, and the time shall be capable of being selected by the operator. Anti-passback shall operate on a system wide basis across multiple NICs and across multiple ports on a single server.

5. The system shall provide for automatic lock/unlock of access-controlled doors on a scheduled basis using time zones.
6. Each card and cardholder shall be entered into the database prior to use. Each card can be manually disabled at any time without the requirement to delete the card. Each card can then be subsequently re-enabled at a later time.
7. Card records shall include the entry of activation and deactivation dates to provide for the automatic enabling and expiring of the card record.
8. The system shall provide the capability of setting a parameter of days whereby cards will be automatically disabled if they are not used at all for access for the preset number of days (i.e., 30 days, 60 days, etc.). Any card can be subsequently re-enabled at any time.
9. The operating mode of access-controlled doors shall be indicated as locked, unlocked, or controlled. The door status shall be indicated as open or closed.
10. The system shall provide for the monitoring of the reader controlled door position in order to detect and report door-forced-open and door-held-open alarm conditions. Door-held-open condition shall be based on a user-adjustable time period. The act of opening the door shall initiate the door timer, and also cause the immediate reset of the door lock.
11. Each cardholder shall be specified with access authority to a combination of up to 128 security areas and 32 groups of security areas, each security area comprised of one or more card reader controlled doors. Up to 20,000 security areas may be defined. Each individual security area or group of security areas designated as authorized to an

individual cardholder shall include a time zone assignment for that specific area. Each cardholder may be specified as authorized access to any, or all, or any combination of the 20,000 security areas.

12. The system shall provide for the designation of certain calendar days to be holidays, with special access privileges and system activity to be specified for those days.
13. The system shall provide the capability to unlock a card reader controlled door and/or mask (shunt) the door alarm, via a request-to-exit door device (motion sensor, exit push button, etc.). The capability shall be software programmable to allow selectable exit reporting.
14. All system controlled electric locks shall be capable of being unlocked via operator command at a workstation.
15. The system shall provide for a completely downloaded and distributed database such that access control decisions are made locally at the access controller and, in the event of the failure of the host computer or loss of communications to the host computer, the access control system shall continue to operate using full database information for all cardholders including security areas authorizations, time zones, expiration dates of cards, holidays, etc. At no time after a card has been entered into the database of the file server and validated, shall the system fail to respond to an access request by a valid cardholder. In the event that the database in the access control field panel is being downloaded or the database is corrupted or voided for any reason, the ACAM server shall make the access transaction decisions based on the current information held in the ACAM server. (Restricted subsets of access control privileges and time zone facilities in the distributed database will not be accepted.)
16. The system shall be capable of utilizing dial-up modems to communicate with remote NICs, so that operators can download cardholder database information to remotely located card reader panels, and upload historical transaction information. The

system shall also be capable of receiving alarm transactions (active alarm, loss of primary power, door forced, etc.) from the remote NICs at any time. The system shall include a provision for specific Alarm Only communication ports, which are to be accessed only for the reporting of alarm information from the NICs. These ports must not be used for upload and download of database information or transaction history. Each NIC must be capable of dialing at least eight telephone numbers in order to communicate with the ACAM system. Failure to communicate on all programmed telephone numbers shall result in a local alarm being generated at the NIC.

2. Alarm Monitor Points:
   a. The security access control system shall provide the following alarm monitoring and reporting functions:
      1. Each supervised security system input point (door contacts, motion detectors and other associated alarm inputs) shall have a user-specified 16 character minimum, text identifier. Each point shall be software programmable as either an alarm or non-alarm event, and shall be capable of reporting active alarm, secure, short circuit trouble, open circuit trouble, ground fault trouble and circuit fault trouble.
      2. The system shall allow masking and unmasking (shunt/enable) of alarm points manually by the operator, automatically by time zone, automatically by the activation of another alarm point, or, where required, by a cardholder from a reader keypad. The system shall not allow an alarm point to be masked if the alarm point is in a trouble condition.
      3. All alarm points shall be individually annunciated upon any change of state. Alarm contacts shall not be connected in parallel or series in zones, unless specifically shown on the contract drawings or stated herein. Double doors with alarm contacts on each leaf of the double door unit may be wired in series, for that double door unit.
      4. All door contacts and request-to-exit devices must be connected in such a manner to provide five-state supervised alarm monitoring. They must be terminated at the RRE module, and shall not require direct

wiring to the NIC. The input point used for door contact shall be user-configurable.

5. The system shall provide for 16 levels of alarm prioritization. The priority determines the order in which a given alarm report will be presented at an ACAM client workstation for disposition when more than one alarm is simultaneously active.
6. The system shall provide for special purpose alarm monitoring and/or transaction reporting for specific events, such as, but not limited to the following:
   a. Duress condition at a card reader.
   b. Anti-passback or tailgating violation.
   c. Rejected access request.
   d. Card reader tamper.
   e. Card reader off-line.
   f. Controller cabinet tamper.
   g. Commercial AC power failure.
   h. Controller communications failure.
   i. Low battery at UPS power supply.
7. All input points may be configured and/or linked to activate a relay output control point or group of control points. Any one input point may be linked to a minimum of at least 64 output control relays.
8. Unacknowledged alarm reports (an alarm condition that has not been acknowledged within a user-specified time period at a client workstation) shall initiate at a designated alternate client workstation on the system.

3. Relay Output Points:
   a. The security access control system shall provide the following relay output control and operational functions:
      1. Each security system output point (door lock, gate controller and other associated relay outputs) shall have a user-specified 16 character minimum, text identifier. Each point shall be software programmable for activation and deactivation.
      2. The system shall allow activation and deactivation of output points manually by the operator, automatically by time zone, automatically by the activation of an alarm point, or, where required, by a card reader.

4. Data Management:
   a. The system shall provide for the following database management capabilities:
      1. The software shall be capable of providing for the recall of system historical transactions with a minimum of one million transactions recallable by operator command from the main event transaction file on the file server hard disk. Additional events may be recalled directly from an archived history log file on a removable hard disk cartridge.
      2. Data searching parameters shall be provided as a menu driven feature of the ACAM system software. The search capability shall include, but is not limited to the following:
         a. Card activity.
         b. Cardholder, by card number or name.
         c. Card readers.
         d. Security areas.
         e. Alarm points.
         f. Alarm categories.
         g. Date and time periods.
      3. The software shall provide report creation capabilities which offer search, organize and sorting according to the operator instructions, and have the ability to print, spool, or display a full report at a printer or client workstation.
      4. All operator commands and database entry functions shall be completely menu driven with plain English text and prompts, and the system shall provide on-screen Help information.
      5. All access to the operator system functions shall require the entry of a valid password. A password must be used by the operator, manager, or administrator to access the system, access authority for each password is completely user-selectable by individual menu selection.

B. Networked Intelligent Controller

1. The Networked Intelligent Controller (NIC) shall be ACUXL16-E1L08A 16 reader controllers w/LAN and power supply. The Networked Intelligent Controllers shall be a

microprocessor-based device, which utilizes a 32-bit processor and a 32-bit bus structure. The controller shall have a minimum clock speed of 90 MHz, and shall be provided with at least 16 Mbytes of battery backed dynamic RAM. The controller shall feature a direct LAN/WAN connection to the controller bus structure in addition to two RS-232 or RS-485 connections, all of which should be designed for use in communication with the access control and alarm monitoring (ACAM) system server. The communication architecture of the NIC shall be such that in the event that the primary communication channel to the ACAM server is lost, the unit shall be capable of automatically switching to a secondary communication channel using one of the host RS-232 or RS-485 connections, and if required shall be able to establish communications via dial-up modem. The NIC shall be provided with a parallel printer port, which will enable it to print transaction data during loss of communication with the ACAM server. The NIC shall be capable of dynamically allocating its memory between database information and transaction history, which shall be stored if the controller has lost communication with the ACAM server. Such transaction history shall be automatically uploaded to the ACAM server once communication has been restored.

2. In its maximum configuration, the NIC shall be capable of storing 500,000 cardholders, and its memory utilization shall be such that if storing database information for 10,000 cardholders, it shall also be capable of storing one million transactions.
3. The NIC shall support the monitoring and control of 16 card readers, with or without keypads. It shall also be provided with at least 12 five-state, fully supervised and fully configurable input points, and at least 12 fully configurable auxiliary output control relays mounted on the main circuit board.
4. Each controller must also be capable of expansion, by external Remote Input Modules (RIMs) and/or Remote Relay Modules (RRMs), to support a combination of up to 172 fully configurable five-state supervised input points or 156 output relays per NIC depending on configuration.
5. Each NIC shall be provided with a UL Listed uninterruptible power supply (UPS) mounted within the NIC enclosure. It shall provide sufficient battery backup to sustain

complete operational effectiveness including Remote Reader Electronic (RRE) modules, card readers, electric locks (fail secure), RIMs and RRMs for a minimum of four hours of normal operation.

6. Each NIC shall utilize on-board self-diagnostic LEDs, removable terminal strips and a pop-in/pop-out circuit board.
7. Each NIC in addition to its on-board LAN/WAN connection shall support RS232 and multi-drop RS-485 communication topologies. Provision of external LAN terminal server devices that are connected though serial communications to the NIC are not acceptable.
8. Each NIC shall support RS-485 bi-directional communication paths (dual multidrop paths back to ACAM file server) with no additional hardware or firmware required.
9. Each NIC shall be supplied with all specified options available, including an enclosure with a tamper switch.
10. Each NIC shall be capable of reporting the following alarm conditions to the ACAM file server:
    a. Enclosure door tamper.
    b. Primary power failure.
    c. Low battery conditions.
    d. Loss of communications.
    e. All access control violations.
11. The quantity and location of NICs shall be as specified in contract documents and drawings.

C. Remote Reader Electronic Modules

1. The Remote Reader Electronic (RRE) modules shall be model RRE-04-E1L. The Remote Reader Electronics shall be provided to support all card readers, door contact switches, request-to-exit devices and electric locks. The RRE modules shall support all industry standard card reader technologies (magnetic stripe, Wiegand, bar code, barium ferrite, and proximity) as well as keypads and compatible biometric devices. These modules shall support the connection of four card devices as required.
2. Each RRE module shall support five-state supervised input points, output relays, and shall provide power outputs of 5 VDC, 12 VDC and 24 VDC output at 500 mA to power card readers, biometric devices, request to exit (REX) devices and

door strikes. Each RRE module shall be capable of being powered by the on-board UPS of a NIC to avoid the need for power supplies and 115-volt outlets to be located near controlled doors. Each RRE shall also be capable of being powered by a local 24 VDC UPS where required.

3. RRE modules shall utilize on-board self-diagnostic LEDs, removable terminal strips and pop-in/pop-out circuit boards.
4. RRE modules shall be supplied with all specified options available, including an enclosure with an enclosure tamper switch.
5. The quantity and location of RRE modules shall be as specified in contract documents and drawings.

D. System Software

1. The access control system software shall be a GE Security Model, Sapphire Pro Series. The software shall be the latest version available from the manufacturer at the time of installation.
2. The software system design shall be object oriented and shall be a native 32-bit application running under the Windows 2000 operating system. All client workstations and the server(s) shall have full system functionality and shall not be segregated in any way by function, except as defined by the user authentications of user name and password.
3. The system shall have a simple, easy to use graphical user interface which is browser based, and all functions shall be accessible by use of either mouse or keyboard. Help text shall be provided for each screen function, and shall be sufficiently interactive that a user may access page help directly and be provided with explicit information relevant to the particular screen being displayed.
4. The system shall have a distributed architecture, however the central server shall have the capability to make transaction decisions for access requests, alarm handling and output control. The software shall be provided with a high-speed real time functionality, which allows the server to take over the transaction handling function of NICs and Video NICs which are being downloaded, or whose database is incomplete or corrupted, and thus maintain the fully

functional access and security response of the NICs under these circumstances. This same real time functionality shall provide for linking of inputs and outputs globally across all NICs within the system on a single ACAM server, and also provide the same global anti-passback linking of card readers across all the NICs connected to a single server.

5. The system shall be provided with the capability to download all of the NICs and Video NICs on the ACAM server system simultaneously. Constraints requiring downloads to NICs in groups is not acceptable.
6. It is vitally important that the access and alarm functionality of the system shall in no way be impaired during periods when database information is being downloaded to NICs or other field devices, or when these NICs or other field devices have insufficient information to make necessary transaction decisions. Thus, it is unacceptable for the performance of NICs to be degraded in any way. Access decisions based solely on company codes or facility codes or even a combination of the two are not acceptable.
7. The system software architecture shall be designed not only to provide a high speed open architecture platform for individual single server applications, but also be specifically designed to insure high speed, high integrity partitioning and redundancy for global, large cardholder database systems.
8. The access control system software shall, as a minimum, support the following features:
   a. Cardholder records—20,000 maximum.
   b. Card readers—512 maximum per server.
   c. Alarm input points—10,000 maximum per server.
   d. Relay outputs 10,000 maximum.
   e. Client workstations—32.
   f. Operator passwords—512.
   g. Up to 2,000 security areas (controlled areas) per server.
   h. 255 time zones (eight time intervals per time zone).
   i. 128 data fields per cardholder.
   j. 16 alarm priorities.
   k. 16 user-defined alarm categories.
   l. 512 action/instruction text messages.
   m. Global and local, hard, soft and timed anti-passback/anti-tailgate capability.

n. Configurable alarm-to-relay linking, downloaded to field controllers for local operation.
o. Configurable automatic time zone controlled commands, downloaded to the field controller for automatic local operation.
p. Configurable automatic, time controlled report generation and/or disk backup commands.
q. Visitor logging and badging utility.
r. History/audit trail.
s. Ability to respond to access requests/alarm conditions before and during download to networked intelligent controllers.
t. Automatic card activation and deactivation.
u. Global and local alarm masking by operator or cardholder.
v. Access activity analysis by card reader.
w. High integrity dial-up capability to support access control panels and readers at remotely located sites via dial-up communication over public switched telephone network, and to download database and upload transaction history based on operator commands. This feature shall provide the capability to segregate dial-up modem ports for Alarm Only reception, and for uploading and downloading database and transaction history information, so that even if all the upload/download ports are in use, the remote dial-up NICs can establish communication through Alarm Only modem ports to annunciate alarm conditions. The memory systems and functionality of the remote NICs shall be such that data corruption during database download, or interrupted communications during download shall not interfere with the ability of the remote NICs to continue to operate normally using database information previously downloaded to it.
x. Capability to define within the system up to 127 variable card formats and have each card reader able to read three separate formats.
y. Integrated video imaging/photo-badging system incorporating a complete multilayer, drag and drop, WYSIWYG, database keyed badge design facility, editor, and drawing package. The system must be capable of allowing enrollment facilities at any designated work-

station, and displaying photo-images of cardholders at any workstation on an individual system.

z. The system shall be capable of interfacing through NICs to keypads or card reader/keypad combinations with LCD displays allowing the system to operate as a proprietary burglar alarm system for designated security areas. The system shall provide delayed alarm reporting and masking facilities using these devices so that arming and disarming of security areas is delayed for a preset period. This will allow personnel entering an area to have the opportunity to disarm (mask) the alarm reporting facility by entering a code into the keypad, entering a card or both. During this entry delay period the LCD/keypad/card reader device shall emit an audible tone until the area is disarmed (masked). The system will allow the user to arm (unmask) the system before leaving the designated area, by entering data on the keypad, using a card or both, and the display shall provide information concerning the status of any unsecured alarm points in the area. Once the security area has been armed, the device will emit an audible tone for a predetermined time, and delay reporting alarm status for this same time to allow the occupant time to exit the secured area.

9. On-Line System Management and Reporting:
   a. The system shall maintain, on disk, an event transaction log file, and be capable of historical data reports as well as cardholder report listings in a variety of formats.
10. System Event Transaction Log File:
   a. The system shall maintain an event transaction log file on hard disk for the recording of all historical event log data.
   b. The historical data file shall maintain the most recent one million event transactions without having to resort to archived media.
      1. The system shall warn the user of the need to archive historical data before data is over-written.
      2. The system shall provide the utilities by which the historical event log file may be backed up to a removable disk cartridge of not less than 1GB capacity, which may be accessed on-line, without the need to copy the archived data back to hard

disk. The system must be capable of recalling historical events directly from the back up magnetic media without the need to interrupt normal on-line activity of the ACAM system.

11. Historical Reports:
    a. The system shall be capable of producing the following reports, based on logged historical events over a specified date and time period, both individually and in any combination.
        1. Report of valid accesses for a selected cardholder, group of cardholders, selected card reader, group of card readers and selected areas.
        2. Report of rejected access attempts for a selected cardholder, group of cardholders, selected card reader, group of card readers and selected areas.
        3. Report of alarm activations for a selected alarm point, group of alarm points, selected category or type of alarms, and by selected areas or group of areas.
        4. Report of alarm acknowledgments for a selected alarm or group of alarms.
        5. Report of operator entered comments in conjunction with alarm acknowledgments.
        6. Report of manual operator override commands such as performed alarm point masking/unmasking, manual card reader door locking and unlocking, and manual auxiliary relay activate/deactivate.
        7. Report of automatic time controlled system commands such as automatic masking/unmasking, and automatic door lock/unlocks.
        8. Report of visitor card valid access and rejected access attempts.
        9. Report of access statistics including the number of valid accesses, rejected access attempts, and card read errors, reported by selected card readers, or group of card readers, or by selected areas, over a selected date and time period.
        10. In addition, the system shall offer the user the option of directing the historical reports to a client workstation color monitor for display or to a report printer.

12. Cardholder Reports
    a. The system shall be capable of producing lists of selected cardholder data records on a client workstation color monitor and/or a report printer. The system shall allow the user to select sorting by card number, cardholder name or other fields.
        1. Standard cardholder record reports may be requested by an operator, with the data records sorted numerically by encoded card number, alphabetically by cardholder name, numerically by cardholder ID number, and numerically by the embossed card serial number. Such listings may also be requested to include only those cardholders who are authorized access to a specified area.
        2. Special Ad Hoc reports may be created by the operator to provide cardholder record listings that include only operator specified data fields. Each report may include conditional testing on up to 16 data fields in order to include data for only those cardholders that comply with those conditions specified. Each report shall be capable of being sorted in alphabetical or numeric order.
        3. Cardholder report formats: The system shall allow the user to create and design the Ad Hoc reports with report format names. The system shall save and store these named formats on the system hard disk for later use and recall by format name.

E. Proximity Readers

1. Proximity card readers shall be HID Corporation ThinLine II model number 5395CK100 (Wiegand, black with pigtail).
2. Provide surface mounting style 125 KHz proximity card readers suitable for wall or US 2-S single-gang box mounting, and for mounting configurations as shown on the project plans.
3. The reader shall be capable of reading access control data in standard Wiegand formats up to 84 bits in length from any HID Proximity card or equivalent, outputting the data in one of the following configurations:
    a. The card reader shall output credential data in compliance with the SIA AC-01 Wiegand standard, compatible with all standard access control systems.

   b. The card reader shall output credential data using a Clock and Data interface, and be compatible with systems requiring a magnetic stripe reader
4. The reader shall be capable of outputting a periodic reader supervision message at a configurable time interval, enabling the host system to signal an alarm condition based on the absence of this message.
5. The Proximity card reader shall provide the ability to change operational features in the field through the use of a factory-programmed command card. Command card operational programming options shall include:
   a. Reader beeps and flashes green on a card read, LED normally red, single line control of LED.
   b. Reader flashes green on a card read, LED normally red, single line control of LED.
   c. Reader beeps on a card read, LED normally red, single line control of LED.
   d. Beeper and LED are controlled by host only, LED normally red, single line control of LED.
   e. Reader beeps and flashes green on a card read, LED normally off, red and green LED's controlled individually.
   f. Reader flashes green on a card read, LED normally off, red and green LEDs controlled individually.
   g. Reader beeps on a card read, LED normally off, red and green LEDs controlled individually.
   h. Beeper and LED are controlled by host only, LED normally off, red and green LED controlled individually
   i. Change from Wiegand to Mag Stripe output format
   j. Change from Mag Stripe to Wiegand output format
   k. Reset to Factory Defaults
6. Proximity card readers shall provide the following programmable audio/visual indication:
   a. A piezoelectric sounder shall provide an audible tone upon successful power up/self test, good card read, or whenever the beeper control line is asserted by the host.
   b. A bi-color, red/green LED shall light upon successful power up/self test, good card read, or whenever the LED control line(s) are asserted by the host.
   c. The reader shall have individual control lines for the sounder, and for red and green LED indication. When

the LED control lines are asserted simultaneously, an amber LED indication will occur.

7. The reader shall have a configurable hold input, which when asserted shall either buffer a single card read or disable the reader, until the line is released. This input may be used for special applications or with loop detectors.
8. The reader shall require that a card, once read, must be removed from the RF field for one second before it will be read again, to prevent multiple reads from a single card presentation and anti-passback errors.
9. Typical proximity card read range shall be up to:
   a. 5.5″ (14 cm) using HID Proxcard II card
   b. 5″ (12.5 cm) using HID ISOProx or DuoProx cards
   c. 2″ (5 cm) using HID ProxKey II key fob
   d. 2.5″ (6.25 cm) using HID Microprox Tag
   e. 5″ (12.5 cm) using HID ***iCLASS*** Prox
   f. 1.5″ (3.8 cm) using HID Prox/Wiegand card
10. Proximity card readers shall meet the following physical specifications:
   a. Dimensions: 4.70 × 3.0 × 0.68″ (11.9 × 7.6 × 1.7 cm)
   b. Weight: 3.3 oz (94 g)
   c. Material: UL94 Polycarbonate
   d. Two-part design with separate reader body and cover.
   e. Color: Black, Gray, White or Beige as approved by the project consultant.
11. Proximity card readers shall meet the following electrical specifications:
   a. Operating voltage: 5–16 VDC, reverse voltage protected. Linear power supply recommended.
   b. Current requirements: (average/peak) 20/115 mA @ 12 VDC
12. Proximity card readers shall meet the following certifications:
   a. UL 294
   b. Canada/UL 294
   c. FCC Certification
   d. Canada Radio Certification
   e. EU and CB Scheme Electrical Safety
   f. U—R&TTE Directive
   g. CE Mark (Europe)

   h. C-Tick (Australia)
   i. New Zealand
   j. Taiwan
   k. Korea
   l. China
13. Proximity card readers shall meet the following environmental specifications:
    a. Operating temperature: –22 to 150 degrees F (–30 to 65 degrees C)
    b. Operating humidity: 0% to 95% relative humidity non-condensing
    c. Weatherized design suitable to withstand harsh environments. The reader shall be of potted, polycarbonate material, sealed to a NEMA rating of 4X (IP55).
14. Proximity card reader cabling requirements shall be:
    a. Cable distance: Wiegand: 500 feet (150 m); Clock & Data: 50 feet (15 m)
    b. Cable type: 5-conductor #22 AWG w/overall shield. Additional conductors will be required for 2-line LED control, beeper, hold, or card present functions
    c. Standard reader termination: 18″ (.5 m) cable pigtail
15. Warranty of Proximity card readers shall be lifetime against defects in materials and workmanship.
16. The quantity and location of readers shall be as specified in contract documents and drawings.

F. Hub

1. The HUB shall be a Linksys model EFAH05W Etherfast 10/100 Auto-Sensing Hub.
2. The Hub shall use standards IEEE 802.3, IEEE 802.3u and protocol CSMA-CD.
3. It shall have 5 10/100 Auto-Sensing RJ-45 ports and 1 10/100 Auto-Sensing Uplink port.
4. The speed of the Hub shall be 10 Mbps (10BaseT0 and 100 Mbps (100BaseTX).
5. The cabling for the Hub shall be Category 3 or 5 UTP for 10BaseT and Category 5 UTP or better for 100BaseTX.
6. The topology of the Hub shall be Star.
7. The Hub shall have power, 10/100 collision, 100 Mbps, link/activity, and SW LEDs.
8. The dimensions of the Hub shall be 7.3″ × 6.1″ × 1.75″.

9. The Hub shall be externally powered by an AC adapter.
10. The Hub shall have an operating temperature of 32°F to 122°F and an operating humidity of 10% to 85%, non-condensing.

G. Electric Strike

1. The electric strikes shall be Von Duprin model 5100 electric strikes. The color of the face plates (aluminum, black or dark brown) shall be determined by the color of each applicable door.
2. The electric strike shall have a die cast aluminum back box, internal steel components, a high density steel keeper that can withstand 1300 lbs. of force and stamped steel faceplates.
3. The electric strikes shall be capable of continuous duty at 12 or 24 volts DC.
4. The electric strikes shall be field adjustable to fail-safe or fail-secure.
5. The electric strike shall have an adjustable keeper with the keeper pocket being 1-7/16″L × ½″D × 11/16″–13/16″W.
6. The electric strikes shall include three faceplates: 4-7/8″ round corner ANSI A115.2, 4-7/8″ square corner ANSI A115.2, and 7-15/16″ long round corner wood frame.
7. The electric strikes shall be non-handed.
8. The electric strike back box dimensions shall be 3-3/8″ × 1″ × 1-11/32″ and overall strike depth shall be 1-11/16″.
9. The 12 V DC version shall draw .38 amps and the 24 V DC version shall draw .19 amps, ±10% @ 77°F.
10. The quantity and location of the electric strikes shall be as specified in the contract documents and drawings.

H. Request To Exit Motion Detector

1. The request to exit motion detectors shall be Bosch model DS150i.
2. The detector shall operate at 12 or 24 V, AC or DC. The current draw shall be 26 mA @ 12 V DC.
3. The detector shall be wall or ceiling mounted for single or double door use.
4. The detector shall have selectable fail-safe and fail-secure modes.

5. The detector shall include 2 form "C" relay contacts.
6. The detector's relay latch time shall be adjustable to 60 seconds.
7. The detector shall have a programmable resettable (accumulative) or non-resettable (counting) timer mode.
8. The dimensions of the detector shall be 1.5″ × 6.25″ × 1.5″ and shall have a high impact ABS plastic enclosure.
9. The detector shall have internal vertical pointability and a wrap-around coverage pattern.
10. The detector shall have one activation LED.
11. The operating temperature of the detector shall be –20°F to 120°F.
12. The quantity and location of the request to exit motion detectors shall be as specified in contract documents and drawings.

I. Flush Mount Door Contacts

1. The flush mount door contacts shall be GE Security model 1076W.
2. The contact shall contain a hermetically sealed magnetic reed switch that shall be potted in the contact housing with a polyurethane based compound.
3. The contact and magnet housing shall snap lock into a 1″ diameter hole.
4. The loop type shall be open or closed with a SPDT electrical configuration.
5. The contact housing shall be molded of flame retardant ABS plastic.
6. The color of the contact shall be determined by the individual door: off-white, mahogany brown, or gray.
7. The quantity and location of the flush mount door contacts shall be as specified in contract documents and drawings.

J. Door Release Button

1. The door release button shall be Alarm Controls model TS-18.
2. It shall be a momentary pushbutton protected by a 1″ diameter guard ring and enclosed in a black plastic box with two mounting ears.
3. The button shall have SPDT contacts, rated 4 amp @ 28 VDC.

4. The button switch shall be terminated with color coded leads.
5. The dimensions of the button shall be 1″H × 1.5″W × 2″L.
6. The quantity and location of the door release button shall be as specified in contract documents and drawings.

2.3 BURGLARY SYSTEM SPECIFICATIONS

A. Control Panel

1. The control panel shall be an Ademco model Vista-50P w/ battery back-up and a 4208U eight zone expansion module.
2. The panel shall have nine hard wired zones standard and be expandable to 87 total zones with eight partitions standard.
3. There shall be 14 zone types to choose from with programmable swinger suppression.
4. The panel shall support up to 75 user codes in all partitions with seven authority levels.
5. The panel shall have four trigger outputs.
6. The panel shall have a 224 event memory log.
7. The panel shall be powered by a 16.5 VAC/40 VA transformer, shall have aux power of 12 VDC, 750 mA maximum, and shall have an alarm output of 12 VDC, 1.7 Amps maximum.
8. The panel's communicator shall support formats Ademco Contact ID, Ademco 4 + 2 Express, and Ademco low speed.
9. The communicator shall have the following: 3 + 1, 4 + 1, and 4 + 2 reporting; split, dual, and split/dual reporting; expanded reporting; true dial tone detection; double pole line seizure; and AC loss and restoral reporting.
10. The panel shall have fast downloading capability.
11. The quantity and location of the control panel shall be as specified in contract documents and drawings.

B. Burglary Keypad

1. The burglary keypad shall be an Ademco 6160 Alpha Display Keypad
2. The keypad shall have soft-touch keys continuously backlit for greater visibility.

3. The keypad shall have a 32 character display. Zones and system events shall be displayed in plain English.
4. The keypad shall have a speaker with audible beeps to indicate system status, entry/exit delay and other alarm situations.
5. The keypad shall have four programmable function keys.
6. The keypad shall have a white removable door.
7. The dimensions of the keypad shall be 5-5/16″H × 7-3/8″W × 1-3/16″D.
8. The standby current of the keypad shall be 40 mA and the activated transmission current shall be 160 mA.
9. The quantity and location of the burglary keypads shall be as specified in contract documents and drawings.

C. Glass Break Detector

1. The glass break detectors shall be IntelliSense model FG-730 glass break detectors.
2. The detector shall detect glass breakage up to 30′ and can mount on the ceiling, opposite wall, adjoining wall or same wall as glass.
3. The detector shall positively detect and confirm breaking glass by analyzing both "flex" (impact) and "audio" (shattering) frequencies. For an alarm to occur the flex signal must be followed by an audio signal within a prescribed time frame.
4. The detector shall operate on 10–14 VDC with a draw of 25 mA as 12 VDC.
5. The detector shall have a Form C alarm relay, 500 mA maximum, 30 VDC maximum.
6. The dimensions of the detector shall be 3-7/8″H × 2-2/5″W × 4/5″D.
7. The operating temperature of the detector shall be 32°–120°F.
8. The quantity and location of the glass break detectors shall be as specified in contract documents and drawings.

D. Dual Technology Motion Detector

1. The dual technology motion detector shall be GE Security model RCR-A.
2. The detector shall have a 35′ range at a mounting height of 7′9″ with range settings of 9′, 18′, 27′ and 35′.

3. The detector shall use a combination of passive infrared (PIR) and microwave technologies. The microwave frequency shall be 5.8 GHz.
4. The detector shall have four distinct coverage zones and the target velocity shall be 0.5 ft/sec to 5 ft/sec.
5. The input voltage of the detector shall be 8.5–18 VDC with 27 mA typical current draw and 35 mA maximum current draw.
6. The detector shall have a Form A relay with a rating of 28 VDC, 100 mA maximum.
7. The dimensions of the detector shall be 2.8″W × 2.3″D × 5.1″H and shall be white in color.
8. The quantity and location of the dual technology motion detectors shall be as specified in contract documents and drawings.

E. Temperature Sensor

1. The temperature sensors shall be Winland Electronics Inc. model EA200 EnviroAlert.
2. The sensor shall have an on-board temperature sensor and a second input for a remote temperature, humidity or water sensor.
3. The input voltage of the sensor shall be 12 VDC @ <120 mA.
4. The operating range of the sensor shall be 32° to 122°F and the Lo and Hi limit adjust range shall be –58° to 299°F.
5. The sensor shall have 2 SPDT relays (configurable) for alarm outputs and 1 SPDT relay (non-configurable) for an auxiliary audible alarm output.
6. The dimensions of the sensor shall be 6″ × 4.75″ × 1.25″ and shall weigh .55 lbs.
7. The quantity and location of the temperature sensors shall be as specified in contract documents and drawings.

F. Panic Button

1. The panic button shall be an Ademco 269 R hold-up button.
2. The panic button shall be a hard-wired hold-up switch with stainless steel cover.
3. The button shall have DPDT contacts and twin screw terminals with EOL resistor splicing terminal.

4. The button shall have a reset key for testing and/or reset of alarm.
5. The button shall have silent operation.
6. The quantity and location of panic buttons shall be as specified in contract documents and drawings.

2.4 CCTV SYSTEM SPECIFICATIONS

A. Fixed Dome Cameras

1. The fixed dome cameras shall be Honeywell Magnaview with a HEV28FC (round flush mount) housing or HEV28RC (round surface mount) housing and a HCGC35 (color, standard resolution) camera.
2. The lens for the camera shall be a HLQAV2 (2.6–6.0 mm vari-focal, auto iris) lens.
3. The impact-resistant camera shall consist of a charged-coupled device (CCD) camera board, lens (varifocal auto iris), enclosure and polycarbonate dome. The camera interface is via 75 ohm coaxial media. The camera enclosure is available for indoor and outdoor applications. The camera, lens and enclosure is fully assembled and tested by the manufacturer. The video system supports either NTSC, EIA, PAL or CCIR signal format.
4. The camera is a ⅓″ solid-state color video camera using an interline transfer, CCD image sensor meeting NTSC (EIA) or PAL (CCIR) signal format. The horizontal resolution is a minimum of 410 TV lines for standard monochrome, 570 for high monochrome, 350 for standard color and 480 for high color. The electronic iris lens has a speed of 1/60–1/100,000 sec. The camera position has a unique 3D-position adjustment allowing for maximum lens rotation for any angle of view required.
5. The input voltage range is 17–32 VAC and 11–16 VDC with a single power supply board; one set of power connections with auto sensing for 24 VAC or 12 VDC is provided. A built-in surge suppression meets a 1.5 kW transient specification. The input power consumption is equal to or less than 4.5 W (Color). V-Phase and Iris level controls are potentiometers accessible from the front of the camera. All cameras with a phase adjustment via potentiometer are capable of AC line lock. The line lock adjustable range of the vertical

phase is +/− 150°. The camera will switch automatically to internal sync mode when 12VDC is applied.

6. The enclosure and polycarbonate dome can withstand a high impact force. The polycarbonate dome is available in full tint; half tint and clear forms. The half tint version has a maximum light loss no greater than 1.0 f-stop. The full tint version does not exceed 2.0 f-stop (used only with monochrome cameras). The dome consists of 3.25″ diameter, 0.125″ thickness that has high optical clarity. The dome is UV-stabilized, suitable for use in areas exposed to sunlight. It is equipped with a hard coat for scratch resistance. The enclosure meets Nema 4X and IP66X and is designed for indoor and outdoor all weather use.
7. For the flush mount version, the camera chassis is fabricated from polycarbonate material. The top enclosure is fabricated from machined aluminum alloy. It also includes captive security lid screws. An adapter plate is also provided for additional support when mounting to ceiling tiles. The dimensions of the camera do not exceed 5.93″ × 3.80″ or 151 mm × 97 mm. The weight of the camera, including both the camera and the lens, is no more than 1.75 lbs (0.79 kg).
8. For the surface mount version, the enclosure is fabricated from cast aluminum alloy and finished with powder white urethane. It includes captive security lid screws and a hinge enabling the lid to remain attached during installation. The enclosure is suitable for surface mounting, with mounting holes enclosed in individual environmental chambers. The dimensions of the camera do not exceed 6.1″ × 3.9″ or 155 mm × 97 mm. The weight of the camera, including both the camera and lens, is no more than 3.0 lbs (1.37 kg). The enclosure can be mounted to a standard 4S electrical box.
9. For the flush mount version, the camera has a cable exit on the back of the enclosure. The cable length supplied is 8″ for power and 6″ for video.
10. For the surface mount version, the camera has a ¾″ NPT standard conduit fitting on the side of the enclosure. The camera also has a ½″ conduit fitting on the rear of the enclosure.
11. The ambient operating temperature shall be 13°F to 122°F.
12. The quantity and location of the cameras shall be as specified in contract documents and drawings.

B. Digital Video Recorders

1. The digital video recorders shall be GE Security model DVMRE-PRO16-600CDRW.
2. The recorder shall support up to sixteen cameras and have a 600 gig hard drive.
3. The Digital Video Multiplexer Recorder with Ethernet connectivity (DVMRe Pro) shall be as manufactured by GE Security. The DVMRe Pro shall require minimal training for the end user. The unit shall operate like a conventional multiplexer and VCR with local display monitors for live and playback viewing while the system continues to record new images. It shall be an integrated security system, capable of time division multiplexing multiple cameras and storing their digitized and compressed images on integral hard disk drives for fast search and retrieval either locally at the unit, or from a remote workstation using a Graphical User Interface (GUI).
4. Additionally, the system shall provide automated alarm handling while operating in normal Triplex mode. Upon receipt of an alarm, the system shall be able to automatically change display and record speed, provide relay output operation, and provide serial data transfer to a host. The system shall be able to determine alarm change of state (COS) conditions from integral motion detection, hard-wired alarm inputs, or serial data transfer from a host. It shall similarly be able to sense an event COS by receipt of text data from a foreign host through a serial port on the unit. During investigations, it shall be possible to search and retrieve stored video data by date, time, camera, alarm, and transaction text, or by selecting a hot spot within a camera's field of view and searching for a change of state.
5. The DVMRe Pro shall include, but not be limited to the following:
   a. The DVMRe Pro shall function as a standalone unit. It shall not require the use of a personal computer, special monitors, or other peripheral devices for either programming or operation. Live and recorded playback of video images shall display on conventional CCTV monitors.
   b. The DVMRe Pro shall be capable of displaying onscreen text and menus in more than one language. This shall be user selectable via the menu system.

c. The DVMRe Pro shall have elastomeric buttons and a jog/shuttle integrated into the front panel of the unit, used for menu navigation, setup, and control of the unit, with no need for an external control device.
d. The DVMRe Pro shall use an easy-to-read, onscreen menu system of pull-down and pop-up selections.
e. The DVMRe Pro shall use a battery to back up memory that stores the time, date, and all internal programming functions.
f. The DVMRe Pro shall have summary setup view screens to show the entire system setup at a glance.
g. The DVMRe Pro shall have a built-in CD-R drive for archiving video as evidence.
h. The DVMRe Pro shall be capable of sending images to an external printer, which shall be activated from a single button on the front panel of the unit.
i. The DVMRe Pro shall support AutoInstall to do the following:
    1. Automatically terminate camera inputs
    2. Automatically detect installed cameras
    3. Automatically detect loss of video sync, with LED and on-screen indicators. If video loss is detected during recording, up to five seconds of buffered fields from the corresponding camera will be recorded regardless of any other record rate settings.
    4. Automatically detect color or black-and-white cameras
    5. Automatically control gain per camera, which shall be adjustable by the user
j. The DVMRe Pro shall prevent unauthorized program tampering through the use of at least four password levels, including:
    1. Total user lockout
    2. Supervisor/installer
    3. User/operator
    4. Ethernet access
k. The DVMRe Pro shall be provided with a UL listed low voltage, AC to DC isolated power supply to prevent susceptibility to power spikes, surges, harmonics, and other common electrical disturbance phenomena associated with the installation environment.

6. The digital recorder shall have the following operational features:
   a. Recording
      1. The DVMRe Pro shall record video on an internal hard disk drive. No videotape or videotape recorders shall be required.
      2. The DVMRe Pro shall support user programmable picture capture rates that can be programmed on a per-camera basis. All cameras shall be programmable to capture images in one of the following operating modes:
         a. Time-lapse (TL)
         b. Event (EVT)
         c. Time-lapse plus event (TL + EVT)
      3. At a minimum, the DVMRe Pro shall support the following image capture rates:
         a. 120 pps/100 pps (NTSC/PAL) in Turbo Recording Mode
         b. 60 pps/50 pps
         c. 30 pps/20 pps
         d. 10 pps
         e. 5.0 pps
         f. 1.0 pps
         g. 0.5 pps
         h. 0.2 pps
      4. The DVMRe Pro shall support an alarm record mode that is user programmable. At a minimum, the DVMRe Pro shall support the following alarm mode image capture rates:
         a. 120 pps/100 pps (NTSC/PAL) in Turbo Recording Mode
         b. 60 pps/50 pps (NTSC/PAL)
         c. 30 pps/20 pps
         d. 10 pps
         e. 5.0 pps
         f. 1.0 pps
         g. 0.5 pps
         h. 0.2 pps
      5. The DVMRe Pro shall allow the user to select whether the hard disk recording should automatically overwrite data (starting with the oldest data first) or if the user must confirm the overwrite

before recording will continue when the hard disk is filled.

6. The DVMRe Pro shall have image quality settings, that are adjustable on a per camera basis by the end user, including the following:
   a. Standard (12.5 KB minimum)
   b. Medium (20 KB minimum)
   c. High (27.5 KB minimum)
7. The DVMRe Pro shall support an image equal to or greater than that of a time-lapse video recorder set at the lowest setting, while maintaining all other performance criteria described in this document.
8. The DVMRe Pro shall support from one to five seconds of pre-alarm recording, maintained in a buffer, and shall append this buffer to the beginning of all recorded alarms. Buffer size shall be user programmable. The DVMRe Pro shall continue to record at the alarm rate until the alarm is reset, times out, or is acknowledged as determined by the alarm menu programming.
9. The DVMRe Pro shall support from one to five seconds of pre-event recording, maintained in a buffer, and shall append this buffer to the beginning of all recorded events. Buffer size shall be user programmable. The DVMRe Pro shall continue to record at the event rate until the programmed event duration (from 0 to 200 seconds) expires. When the DVMRe Pro is recording in Turbo Mode, pre-alarm storage shall be suspended.
10. The DVMRe Pro shall allow the user to manually or automatically customize the record rates per camera for events, activity detection, and alarms.
11. The user shall be able to play back images smoothly at normal or fast speeds and in forward or reverse modes, without distortion.
12. The unit shall provide full media search capabilities for archiving, restoring, and playback operations. Search capabilities shall include filters for start/stop times, start/stop dates, alarm and event occurrences, inserted text, and camera number.
13. The unit shall use Wavelet technology to compress and store pictures prior to recording.

14. The DVMRe Pro shall support the recording of all images with a digital watermark. The verification of watermarked images shall reside solely with the manufacturer.

b. Archiving
   1. The DVMRe Pro shall have a built-in CD-W drive
   2. The built in CD-W drive shall create evidence capture CDs capable of being played on a PC without the need for special software.
   3. Recording to the CD-drive shall be operated from the menu system of the unit, and can also be activated using a button on the front panel for capturing evidence during times of duress.

c. Multiplexing
   1. The DVMRe Pro shall be a triplex type unit, allowing simultaneous recording, playback, and live multiscreen viewing at the unit, with no need for additional hardware. When placed in turbo mode, the system shall cease operation and dedicate all internal resources to provide record rates of up to 120 pps.
   2. The DVMRe Pro shall provide the following displays in live, Triplex mode: full screen, sequencing, picture-in-picture, 4-way, 6-way, 7-way, 9-way, 10-way, 13-way, or 16-way.
   3. The DVMRe Pro shall provide the following Triplex displays in playback mode: full screen, sequencing, picture-in-picture, 4-way, 6-way, 7-way, 9-way, 10-way, 13-way, or 16-way.
   4. The DVMRe Pro shall remember the last multiscreen display selected and recall it when switching between full screen and multiscreen views.
   5. The DVMRe Pro shall allow the user to rearrange cameras in any multiscreen display in both live and playback modes.
   6. The DVMRe Pro shall provide one-touch image freezing. An on-screen indication shall be displayed when the image freezing feature is in use.
   7. The DVMRe Pro shall provide one-touch electronic zooming with both LED and on-screen indicators. All areas of zoomed images shall be capable of being viewed through the use of digital pan and tilt functions.

8. The DVMRe Pro shall have a covert camera feature, which removes the selected camera from the live display but allows it to be recorded and viewed during playback by an authorized user. More than one camera can be selected for covert recording.
9. The DVMRe Pro shall incorporate the following display options:
    a. Camera titling with a minimum of up to 12 alphanumeric characters
    b. Title display enable/disable, per monitor
    c. Time/date formatting
    d. Time/date enable/disable, per monitor
10. The DVMRe Pro shall provide image update rates for live and record modes of up to 120 unique pictures per second for NTSC or up to 100 unique pictures per second for PAL.
11. The DVMRe Pro shall support three monitor output modes of operation.
    a. In the single-monitor operation mode, the "main" monitor shall be used for all multiscreen displays in both playback and record. During playback, the main screen shall display both live and playback images. The second "spot" monitor shall be available for displaying analog, full-screen pictures of normal or alarmed cameras.
    b. In the dual-monitor operation mode, the unit shall support simultaneous multiscreen display of live images on the "main" monitor, and the second monitor output shall be used for search operations and the multiscreen display of playback images. The second multiscreen shall also display live multiscreen while not in playback mode.
    c. When in turbo mode, Monitor A will drive a message on the screen, directing the user to watch monitor B. Monitor B shall allow full camera call up of sequencing.

d. Video motion detection
    1. The DVMRe Pro shall support the following two types of user selected video motion detection, with on-screen indications when motion is occurring:

a. Intrusion detection, which shall be programmable by the user to be treated as an alarm.
    1. The DVMRe Pro shall support an on-screen setup scale to determine the optimum sensitivity setting for each camera input.
    2. The DVMRe Pro shall have 256 zones per camera, arranged in a 16 by 16 grid.
    3. The DVMRe Pro shall have 10 levels of sensitivity.
    4. The DVMRe Pro shall have 256 levels of gray per zone.
    5. The DVMRe Pro shall have three levels of false-alarm-rejection processing.
    6. The DVMRe Pro shall have 256 levels for size discrimination.
b. Activity detection, which shall be user programmable to be treated as an event.
c. The DVMRe Pro shall provide a programmable recording priority capability for activity detection that is not to be treated as an event, on a per-camera basis.

2. The DVMRe Pro shall have a motion search feature for locating recorded occurrences of motion detection. It shall be possible during a recorded video search to call up a camera, highlight a hot spot area within the field of view that requires investigation and find all video files that have shown a disturbance in that field of view such as a person passing through the area, an object removed from the area, or an object placed in the area.

e. Alarms
    1. The DVMRe Pro shall support up to 16 alarm inputs (one per camera), programmable as normally open or normally closed from within the menus.
    2. The DVMRe Pro shall support two form-C relays as alarm outputs, each programmable as normally open or normally closed from within the menus, and rated for 0.5 A continuous, 1.0 A momentary. Upon alarm, the system shall be able to execute a change of state (COS) to relay number 1, relay number 2, or both.

3. The DVMRe Pro shall have a fully programmable additional audible device to alert the user to alarms, intrusion detection, and video loss occurrences.
4. The DVMRe Pro shall support alarm latching with three settings, which shall be programmable from the menus as follows:
   a. Latched—the alarm must be reset by an operator, such as for a high priority alarm like a panic alarm or hold-up button.
   b. Transparent—the alarm shall automatically reset when the COS is restored back to its normal condition, such as when tracking an individual from camera to camera through a facility using motion detection, Passive Infrared (PIR), or other sensing devices.
   c. Timed out—the alarm shall automatically reset after a user-defined elapsed time.
5. The DVMRe Pro shall support alarm recording with the user's preference of automatic priority control as follows:
   a. Interleaved—the alarmed camera shall be recorded in every other field as the unit continues to multiplex and record all cameras. Playback emulates near real time with a very high frame rate.
   b. Exclusive—the alarmed camera only shall be recording, producing the fastest possible alarm recording rate.
   c. None—no change shall be made to the recording sequence as the result of an incoming alarm.
6. The DVMRe Pro shall have automatic full screen or programmable associated multicamera, multiscreen alarm displays, that shall change as incoming alarms continue to arrive. As additional alarms arrive, the display monitor shall increase the number of views shown on one screen, with alarmed cameras showing at the top of the screen. It shall be possible, using the telemetry preset control described elsewhere in this specification, to utilize presets with associated alarm display to show the alarmed scene and surrounding escape paths during a high level alarm condition.

7. The DVMRe Pro shall provide status relays that shall link to alarms, motion detection, and video loss.
8. The DVMRe Pro shall provide the ability for the operator to manually activate alarms.
9. The DVMRe Pro shall have an alarm history display capable of showing the last 100 alarms received by the system.
10. The DVMRe Pro shall be supplied with an adapter card with screw-terminal connections to facilitate easy connection of alarms and other input/output signals.
11. The DVMRe Pro shall have a connection for an external alarm acknowledge/silence switch.

f. Macro programming
   1. The DVMRe Pro shall support up to 16 user-programmable "macro" functions (with 32-keystroke memory) for simple two-key operation of multiple features and programming changes. These macros shall be used to record a series of keystrokes, remember them, and execute them when called to by either an alarm association, a timed event, or by manual call up.
   2. The DVMRe Pro shall support up to 20 scheduled events that can be linked to any macro for automatic execution of any programmable macro function.
   3. The DVMRe Pro shall contain a built-in macro for easy switching between standard and daylight savings time.

g. Sub-macro programming
   1. The DVMRe Pro shall support the use of embedded sub-macros through an RS-232 port to allow the unit to send data and commands to other systems, such as cross-point matrix switching systems, credentialized access control systems, or facility management systems.

h. RS-485 communications and networking
   1. The DVMRe Pro shall provide a RS-485 bus and shall support RS-485 networking and control to facilitate operation of the following:
      a. Remote control of system operation, setup, uploading and downloading, and system programming operations

b. Motorized PTZ control
c. Programmable presets on alarm
d. Any combination of up to 32 keypads and digital recorder units
e. Up to 992 PTZ camera receivers
f. Master/slave timekeeping operations
g. System integration

i. RS-232 communications
1. The DVMRe Pro shall support RS-232 communications and control to facilitate:
   a. Remote control of system operation, setup, uploading and downloading, and system programming operations
   b. ASCII output of command strings and data generated by sub-macros
   c. Text insertion shall be supported as follows:
      1. The DVMRe Pro shall accept up to 500 bytes of text with every field.
      2. Each message shall be associated with a single camera.
      3. Four types of event text messages shall be supported:
         a. Start of event (event mode is started and optional text string is stored with first event field)
         b. End of event (event mode is stopped and optional text string is stored with next event field)
         c. Event snapshot (at least one field from the event camera is recorded with an optional text string)
         d. No change (text is added to next field of the selected camera without changing the camera's record rate)
      4. Text messages shall be discarded if the DVMRe Pro is not in record mode.
      5. In full screen playback on the DVMRe Pro, the text will be displayed on the field it was recorded with. Only the first 50 characters shall be displayed in a single line.
      6. The color of the text shall be the same as the color of the camera title (black, white, or

gray) and shall change to green at the instant that that transaction was captured.

7. The text shall be cleared when the play clock changes by five seconds.
8. Text shall not be updated more than once per second unless the frame advance/reverse key is being pressed.

d. The system shall be upgraded through flash programming upgrades of software, using either an RS-232 port or TCP/IP.

e. The manufacturer shall utilize open protocol and protocols shall be supplied as a part of the standard documentation or available elsewhere from the manufacturer.

j. Ethernet communications

1. The DVMRe Pro shall support LAN/WAN Ethernet access.
2. The DVMRe Pro shall support Ethernet bandwidths of 10 MB or 100 MB.
3. The DVMRe Pro shall support simultaneous Ethernet access by not less than two workstations connected to the LAN/WAN.
4. The DVMRe Pro shall be provided with a Graphical User Interface (GUI) software for remote playback and viewing that shall support the Windows 98/NT/2000 operating systems and full searching capabilities. It shall be possible to remotely set up the DVMRe Pro unit using the remote viewing software, save the setup configuration using the Windows™ "Save As" function and then use the data to restore or back up existing sites, or clone new sites that have identical setup configurations.
5. The DVMRe Pro shall not stop recording during any Ethernet access, nor shall it be possible to remotely issue a command via Ethernet to stop the recording.
6. The DVMRe Pro shall allow the user to disable all Ethernet access from the menus.
7. The DVMRe Pro shall allow the user full programming of Ethernet parameters, including the following:

   a. IP address
   b. Default gateway
   c. Sub-net mask

k. PSTN telephone line communications
   1. The DVMRe Pro shall support dial-up access to live and stored images via a standard, manufacturer approved modem connection.
   2. The DVMRe Pro shall not stop recording during any modem access.
   3. The DVMRe Pro shall be provided with remote playback and viewing software that shall support the Windows 98/NT/2000 operating systems and restricted searching capabilities.

7. The DVMRe Pro shall have the following additional specifications:
   a. Video
      1. Total available video memory shall be at least 64 MB.
         a. Live/playback display memory shall be 32 MB.
         b. Record memory shall be 32 MB.
      2. Video sampling rate shall be 27 MHz.
      3. Available colors shall be to specification YUV 4:2:2, providing up to 16.8 million colors.
      4. There shall be 256 grayscale levels.
      5. Horizontal resolution shall be 720 pixels.
      6. Vertical resolution shall be:
         a. 484 active lines NTSC/EIA
         b. 576 active lines PAL/CCIR
      7. Inputs
         a. Camera
            1. There shall be 16 looping camera inputs.
            2. Inputs shall use BNC connectors.
            3. Inputs shall be NTSC/EIA or PAL/CCIR compatible, depending on the model selected.
         b. Signal conditioning
            1. All inputs shall have automatic gain control.
            2. Inputs shall accept video levels from 0.7 to 2.0 V p-p.
      8. Outputs
         a. In Triplex mode, the DVMRe Pro shall have two monitor outputs as follows:
            1. One digital multiscreen monitor-A output
               a. Composite video, BNC connector
               b. NTSC/EIA or PAL/CCIR compatible

2. One multiscreen monitor-B output
   a. Composite video, BNC connector
   b. NTSC/EIA or PAL/CCIR compatible

b. In Turbo Mode, the DVMRe Pro shall have one monitor output as follows:
   1. Monitor A shall display a message stating the system is in Turbo Mode
      a. Composite video, BNC connector
      b. NTSC/EIA or PAL/CCIR compatible
   2. One full-screen monitor-B output
      a. Composite video, BNC connector
      b. NTSC/EIA or PAL/CCIR compatible

b. Audio
   1. The DVMRe Pro shall have one channel of audio that may be enabled or disabled.

c. The DVMRe Pro shall have two RS-232 DB-9 serial data ports to support the following functionality:
   1. Remote communications
   2. Modem communications
   3. Transaction Text insertion

d. The DVMRe Pro shall have one DB-25 multipurpose input/output connector to support the following functionality:
   1. Alarm inputs
   2. Alarm acknowledgement
   3. Relay outputs

e. Electrical
   1. Input voltage: 12 VDC
   2. Power adapter: 90 to 264 VAC/VDC (included)
   3. Power: 35 W nominal

f. Environmental
   1. Operating temperature range: 32 to 104°F (0 to 40°C)
   2. Relative humidity: 90% non-condensing

8. The digital recorder shall conform to these internationally recognized compliance standards:
   a. FCC Part 15
   b. CE
   c. UL 1950
   d. CSA 22.2

9. The quantity and location of digital recorders shall be as specified in contract documents and drawings.

C. Software

1. The remote digital video software shall be the latest version of GE Security WaveReader software
2. The software shall have full or multiscreen live viewing and playback modes.
3. It shall utilize LAN/WAN with bandwidth throttle or dial-up.
4. Video shall be found by time/date range, camera, event, and transaction text.
5. There shall be the ability to view video on the PC while recording.
6. The software shall have an address book that accommodates up to 10,000 sites, with 18 data fields for recording (site name, address, city, state, country, etc.). Just click on any column to sort by the corresponding heading.
7. Security rights shall enable to customize access and control privileges.
8. It shall save, enhance, export, and print images using the built-in WaveStudio feature.
9. It shall create self-contained evidence CDs with QuickWave, which will play on the PC without additional software.
10. The WaveWatch program shall automatically verify digital system operation.

D. Monitor

1. The monitors shall be GE Security model MVC-17HS 17″ color monitors.
2. The monitors shall have a CRT screen with 17-inch diagonal viewing area and be contained in a metal cabinet.
3. The monitor shall meet or exceed the following specifications:
   a. The monitor shall incorporate a 17-inch color CRT screen.
   b. The monitor shall have a horizontal resolution of at least 800 television lines.
   c. The monitor shall have a dual scanning system that is NTSC and PAL compatible.
      1. The scanning frequency shall be 15.734 KHz/60 Hz NTSC and 15.625 KHz/50 Hz PAL.

   d. The monitor shall have a metal cabinet with a built-in speaker and ergonomic handles on each side.
   e. The monitor shall have built-in microprocessor digital control.
   f. The monitor shall use the following signal input:
      1. Composite video 1.0 V p-p at 75 ohms
   g. The monitor shall use the following video connections:
      1. BNC in/out—2 sets
      2. Y/C in/out
   h. The monitor shall use the following audio connections:
      1. RCA in/out—3 sets
   i. The monitor shall have the following front-panel user controls:
      1. Power on/off
      2. Brightness
      3. Contrast
      4. OSD
      5. Enter
   j. The monitor shall have automatic loop-through, self-terminating outputs.
4. The electrical specifications for the monitor shall be as follows:
   a. Power input shall be 100 to 240 VAC, 50/60 Hz
   b. Power consumption shall be 100 W maximum.
5. The environmental specifications for the monitor shall be as follows:
   a. Operating temperature shall be 14 to 122°F (–10 to 50°C).
   b. Operating humidity shall be 10 to 90 percent.
6. The physical specifications for the monitor shall be as follows:
   a. Net weight shall be 44.1 pounds or 20 kilograms.
   b. Dimensions shall be 15.8 (H) × 16.4 (W) × 16.4 (D) inches (400 × 416 × 417 mm).
7. The monitor shall have the following accessories for rack mounting:
   a. KTM-RK-17C shall mount one 17-inch monitor in a 19-inch rack space.
8. The monitor shall conform to these internationally recognized compliance standards:
   a. FCC
   b. UL
   c. C-Tick

9. The quantity and location of the monitors shall be as specified in the contract documents and drawings.

E. Power Supply
   1. The power supplies shall be an Altronix model ALTV2416 CCTV power supply.
   2. The power supply shall have sixteen outputs with a total output current of 8 amps (200 VA) @ 24 VAC or 7 amps (200 VA) @ 28 VAC.
   3. The power supply shall have fuse protected outputs. The output fuses are rated @ 3.5 amps.
   4. The input shall be 115 VAC, 50/60 Hz, 1.9 amps.
   5. The main fuse shall be rated @ 10 amp/250 V. There shall be an illuminated power disconnect circuit breaker with manual reset.
   6. The power supply shall have a Power ON/OFF switch, an AC power LED indicator and surge suppression.
   7. The power supply shall include a gray enclosure and two transformers.
   8. The power supply shall maintain camera synchronization.
   9. The dimensions of the power supply shall be 8.5″ H × 7.5″ W × 3.5″ D and shall have ½″ and ¾″ combination knockouts.
   10. The quantity and location of power supplies shall be as specified in contract documents and drawings.

PART III EXECUTION

3.1 ENGINEERING AND DESIGN

A. The Contractor shall be responsible for fully engineering all final aspects of the project to reflect the intent of the specifications and satisfy the Owner's requirements. The Contractor shall be responsible for providing factory engineering and final review of all application design requirements. The Contractor shall be responsible for all costs related to factory engineering, design review and field testing.

B. The Contractor is required to read the Specifications covering all aspects of the Work and will be held responsible for coordination of the Work performed.

C. The Contractor shall be responsible to provide, at no additional cost, any equipment beyond that shown on drawings or identified in the specifications to provide for a complete system.

D. Should the Contractor have any questions concerning the plans and specifications, they are to contact the Consultant in writing prior to the bid.

E. The Contractor shall coordinate with the Owner for the installation of the CPU, electric lockset, voltage monitor, door hardware including electric hinges and door boring, conduit for readers and electrical power as well as supplying the access cards.

3.2 TESTING AND INSPECTION

A. All systems shall be thoroughly tested upon completion. The Contractor and Consultant shall coordinate with each other to develop an appropriate test report prior to testing. The Contractor shall advise the Consultant and Owner in writing a minimum of 10 business days in advance of the testing so Consultant may witness the same. The Contractor shall provide a written report of test results to the Consultant.

B. Upon completion of the test and test result review, the Consultant will provide a punch list if necessary. The Contractor shall re-test and provide any additional testing documentation requested by the Consultant as many times as necessary until all punch list items are satisfied.

C. All costs associated with the repair of equipment provided by the Contractor and found faulty in the test process shall be the responsibility of the Contractor.

D. Field "checkout" and Owner training shall not commence until satisfactory review of the testing documentation requested.

E. The system testing shall consist of the activation or usage of all field installed devices.

F. System testing shall clearly demonstrate the operation of all elements of the systems.

G. The contractor shall have "as-built" drawings prepared and ready for review at the time of inspection by the Consultant.

3.3 CONDUIT AND CABLING

A. All necessary wiring shall be furnished and installed by the Contractor. All cabling shall be in conformance with the Specifications and manufacturer's recommendations, be U.L. listed, and installed per prevailing electrical codes.

B. All cable shall be plenum rated.

C. In general, all "home run" field wiring may be run in an open manner above hung and drywall ceilings. All cabling shall be tagged, bundled, and supported per prevailing electrical codes. Proposed cabling routes will be fully indicated on the specified shop drawings.

D. All cabling that cannot be installed above ceiling shall be installed within conduit. Minimum conduit sizing shall be ¾". Conduit fill shall not exceed 40% of interior cross sectional area where three or more cables are contained within a single conduit. Conduits will be installed in a manner as to eliminate the potential of obtaining a "hand hold."

E. Cables requiring shields will be installed in such a manner to insure that the shield does not touch the terminal connections. Insulate the shields at the device and panel ends with shrink tubing to completely cover the shield. Provide "one end" grounds on all shields using drain wires. Drain wires will be connected to a dedicated RF ground via a #8 AWG copper cable. Do not connect grounds to a building electrical ground.

F. Lay out conduit and cable runs prior to installation and maintain 1′ clearance of parallel runs between 120 VAC and low voltage conduit and wiring. Cross 120 VAC and low voltage conduit runs at 90° to reduce EMI and RF induction in security wiring.

G. All wires within cables shall be color coded to provide separate identification of the system they are associated with. Wires shall be similarly coded for each system throughout the entire installation.

H. All cables shall be tagged at both ends of all circuits with Brady type markers. All tagging information shall be included on the "as built" drawings. Patch cords between equipment shall be labeled to create as uniform a tagging system as possible. Marking cables with felt tip markers or other markers is not permitted.

I. No intermediate cable splices are allowed without the specific written approval of the Consultant. Request clarification and approval prior to installation of any splices.

J. All wiring in panels shall be neatly dressed and provided with adequate slack for future maintenance and service terminations.

K. All unused wire conductors shall be insulated to eliminate shorts and grounds and tagged as "spare."

L. All wiring or cable shall be fully tested and be free of opens or shorts. All wiring shall be test free of grounds with the exception of circuits that are intended to be connected to the ground-side of protective circuits.

M. Paralleling of multi-conductor cables to create a larger capacity cable is prohibited.

N. All wire and cables entering equipment cabinets and enclosures shall be grouped and tied inside the enclosures on 6-inch centers with self locking nylon cable ties. All wiring shall be grouped in an orderly fashion. Under no circumstances is the use of adhesive tapes (electrical or other) permitted for either permanent or temporary ties or wire management.

O. All wire and cables on the backboard for the controllers and other equipment shall be enclosed within a wire management system. The wire management system shall include cable chases w/covers of appropriate size (example—3″ × 3″ Panduit w/covers).

3.4 OPERATIONAL OBJECTIVES

A. Access Control System

1. When an authorized proximity card is held up to a reader at the required distance, the associated electric strike or electric lockset shall release to unlock the door.

2. The door hardware shall be as such that allows free exit by using the normal door lever operation. The exception to this is the four doors that have readers on both sides. For these doors, exit is by reader also.
3. The request to exit motion detectors used on some doors are not for allowing free exit but rather to temporarily shunt the door during the hours the burglar alarm system is armed so no "forced door" signal is sent to the burglar panel.
4. The doors shall be set up with "propped open" alarms to the PC and the doors specified on the drawings will feed through to the burglar alarm system for "forced open" alarms.
5. The access control system shall be able to be programmed by the Owner through installed software on the Owner supplied PC. Access can be authorized or denied by card, door, time, day or any combination.
6. The glass break detectors and temperature sensors shall be set up for alarms to the PC and feed through to the burglar alarm system for alarms.
7. The receptionist shall use the door release button to unlock the interior reception door.
8. The front reception door will automatically lock and unlock at set times. If an authorized person has not entered the premises in a set time before the door is supposed to unlock, the door will not unlock until an authorized person has entered.

B. Burglar Alarm System

1. The burglar alarm system shall auto arm and disarm at set times. The system will never be armed or disarmed via a keypad.
2. The panic button, temperature sensors, motion detector, glass break detectors and voltage monitor shall be armed 24 hours a day.
3. The burglar keypad shall operate the motion detector. When someone enters check storage room, they will disarm the motion detector via the keypad. When they leave the room they will arm the motion detector via the keypad.
4. If an alarm occurs, the burglar alarm panel shall send a signal to a central monitoring station for dispatch.

C. Camera System

1. The fixed cameras shall be viewed using two monitors.
2. The camera images shall be recorded on two digital recorders. The control of the recorders will be from front panel controls only.
3. Software shall be available for the digital recorder but shall not be installed on a PC.
4. Archiving of camera images shall be done on a CD/DVD drive within the recorders.

3.5 SCOPE OF WORK—SPECIFIC ACCESS CONTROL SYSTEM INSTALLATION CRITERIA

A. Controllers

1. The controllers and reader modules shall be installed per the specifications and manufacturers' instructions.
2. The controllers, reader modules, burglar control panel, CCTV power supplies, and Hub shall be wall mounted on the Owner supplied backboard as shown on the drawings. The Contractor shall use a wire management system in between all the cabinets. The wire management system shall include cable chases w/covers of appropriate size (example—3″ × 3″ Panduit w/covers).
3. The Contractor shall program the system to the Owner's specific requirements. The Owner shall provide the proximity access cards.
4. Connect to Hub per manufacturers' instructions in order to simulate a LAN connection.
5. Connect to card readers, door contacts, request to exit devices, electric locks or strikes, and door release button per manufacturer instructions and these specifications.
6. Specific doors as shown on the drawings will output to the burglary panel for "forced open."

B. Software

1. The Contractor shall install the Sapphire Pro software on the central processing unit.
2. The Contractor shall program the software and load data bases and produce initial reports per the custom requirements as developed by the Owner.

3. The software shall be programmed with the "first person in" feature so the front reception door is not unlocked unless an authorized person has entered the premises during a set time.
4. The Contractor shall be responsible for backing up all system data programmed until the Owner accepts the project as complete. If any data is lost prior to this time, the Contractor shall be responsible for restoring the data.

C. Proximity Readers

1. The readers shall be installed per the specifications, manufacturers' instructions, and ADA height requirements.
2. Install adjacent to door to be controlled.
3. Seal all readers with silicone to eliminate water or moisture ingress.
4. The Owner shall stub up conduit for each reader for installation.

D. Electric Strikes

1. The electric strikes shall be installed per the specifications and manufacturers' instructions.
2. The electric strikes shall be fail-secure.
3. For the double doors, the Owner shall be responsible for supplying and installing the necessary electric hinges and to bore the doors for wiring to the electric strike.

E. Request To Exit Motion Detectors

1. The request to exit motion detectors shall be installed per the specifications and manufacturers' instructions.
2. Mount request for exit motion detector on wall, centerline of door, approximately 6–12" above door frame to centerline of device. Insure detection pattern only "sees" exiting person within 3' radius of door opening.

F. Door Release Button

1. The door release button shall be installed per the specifications and manufacturers' instructions.
2. The door release button shall be installed at the receptionist's desk. Coordinate with Owner for exact mounting location.

G. Door Contacts

1. The door contacts shall be installed per the specifications and manufacturer's instructions.
2. Flush mount door contacts shall be installed at each access controlled door for "door ajar" and "forced open" alarming and wired into the controller.
3. The contacts shall be installed on the top non-hinged side of the door. There shall be no exposed wiring.

3.6 SCOPE OF WORK—SPECIFIC BURGLAR ALARM INSTALLATION CRITERIA

A. Control Panel

1. The control panel and expansion board shall be installed per the specifications and manufacturer's instructions.
2. The control panel shall be programmed for central station monitoring. Coordinate with Owner for required pass codes, dispatch directions, etc. An RJ-31X jack shall be utilized to tie into the existing phone service.
3. The control panel shall be programmed for auto arming and unarming. Coordinate with Owner for specific times and days for this programming.
4. The relay contacts of an Owner supplied voltage monitor shall be wired into a control panel zone as shown on the drawings.

B. Burglar Keypad

1. The burglar keypad shall be installed per the specifications and manufacturer's instructions.
2. The burglary keypad shall be installed at a height of 48″ to the top.
3. The burglary keypad shall control the motion detector only.

C. Glass Break Detectors

1. The glass break detectors shall be installed per the specifications and manufacturer's instructions.
2. The glass break detectors shall be installed on the ceiling within 30′ of all protected glass.

3. Glass break detector mounting locations shall have adequate physical blocking to insure strong and secure attachment.
4. The glass break detector shall be set up as "non-latching."
5. The glass break detectors shall feed through the access control panel before going to the burglar panel so glass break alarms also show up on the Sapphire Pro software.

D. Dual Technology Motion Detectors

1. The dual technology motion detectors shall be installed per specifications and manufacturer's instructions.
2. The motion detectors shall be installed at a height of 96" to the top.
3. The motion detector shall be controlled by the burglar keypad.

E. Temperature Sensors

1. The temperature sensors shall be installed per specifications and manufacturer's instructions.
2. The high heat setting shall be between 85° and 90° F and the low setting shall be 32° F.
3. The temperature sensors shall feed through the access control panel before going to the burglar panel so temperature alarms will also show up on the Sapphire Pro software.

F. Panic Button

1. The panic button shall be installed per specifications and manufacturer's instructions.
2. The panic button shall be installed under the receptionist's desk. Coordinate with Owner for exact mounting location

3.7 SCOPE OF WORK—SPECIFIC CCTV SYSTEM INSTALLATION CRITERIA

A. Fixed Dome Cameras

1. The fixed dome cameras shall be installed per the specifications and manufacturers' instructions.
2. Camera locations are shown as approximate locations on the project drawings. Coordinate with Owner to determine locations prior to the start of installation.

3. Furnish and install dedicated home run power and video cables to each camera location shown on the drawings.
4. The video derived from the vari-focal lens of the cameras shall be approved by the Owner. The Contractor shall be responsible for any final adjustments requested.
5. Connectors used to connect coaxial cable to system equipment shall be of a type that incorporates an integral radiation suppressing sleeve installed with a purpose specific crimping tool. Care shall be taken to insure that the cable is properly prepared and installed to utilize the characteristics of the sleeve. Screw on BNC connectors shall not be utilized. Insure all connectors are tightened firmly together and test for video signal distortion caused by loose connections or terminators.
6. Camera domes are to mount on ceilings requiring special attention to placement. Provide structural blocking as required to insure a system free of video jitter due to vibration or movement of dome assembly to building connection.
7. While no flush mount dome cameras shall be mounted on walls, it is possible one or more of the three surface mounted dome cameras in the dock area will mount on a wall. Coordinate with Owner to verify camera placement.

B. Digital Video Recording System

1. The digital video recording systems shall be installed per the specifications and manufacturers' instructions.
2. The Contractor shall program the system to the Owner's specific requirements.
3. The digital recorder shall record for a minimum of 30 days before writing over itself.
4. The digital video recorders shall be installed on a shelf or desk supplied by the Owner.
5. The Wave reader software shall be given to the Owner but not installed on a PC.
6. Connectors used to connect coaxial cable to system equipment shall be of a type that incorporates an integral radiation suppressing sleeve installed with a purpose specific crimping tool. Care shall be taken to insure that the cable is properly prepared and installed to utilize the characteristics of the sleeve. Screw on BNC connectors shall not be utilized. Insure all connectors are tightened firmly together and test for

video signal distortion caused by loose connections or terminators.

C. Color Video Monitors

1. The color video monitors shall be installed per the specifications and manufacturers' instructions.
2. The monitors shall be installed on a shelf or desk supplied by the Owner.

D. Power Supply

1. The power supplies shall be installed per the specifications, manufacturers' instructions, and prevailing electrical codes.
2. Power supplies shall be hard wired into 120 VAC source by the Owner.
3. No more than one camera will be connected to each of the sixteen fused outputs of each power supply. Do not exceed 90% of overall power supply power output rating.
4. All electrical connections will be made with appropriately sized UL approved cable connections installed with a UL approved crimping tool.
5. Battery back-up shall be 8 hour minimum at full load.

3.8 CUTTING AND PATCHING

A. All cutting and patching required for equipment included in these specifications shall be performed in accordance with all applicable codes and standards.

B. All holes cut through concrete slabs shall be core drilled. No structural members shall be cut without the approval of the Owner.

C. All sleeves through rated walls or floors (sound or fire rated) shall be sealed/packed with an approved and listed-for-purpose material in accordance with manufacturers' instructions, to maintain the fire and/or sound rating of the penetrated wall or floor.

D. Connections to fireproofed steel shall remove fireproof coating only sufficient to make connection. Repair any damaged fireproofing to like new condition per fireproofing specifications.

3.9 MANUALS AND SOFTWARE

A. The Contractor shall provide the Owner two complete sets of operation, maintenance and service manuals for all equipment provided under this contract. The manuals shall be compiled, assembled and indexed in an easily identifiable hardcover form. They shall include the following:

  1. Complete operating instructions.
  2. Complete maintenance instructions, wiring diagrams, troubleshooting instructions.
  3. Preventative maintenance requirements.
  4. Complete parts list for all equipment installed.
  5. Complete collection of manufacturers' specifications.
  6. Manufacturers' warranties.
  7. Software instructions.

B. The Contractor shall supply the Owner a complete set of software CDs for each piece of software installed.

C. The Contractor shall supply the Owner and Consultant with the "as-built" drawings as specified herein.

Document D-1 **Sample Design Specifications**

Appendix E
Sample Design Drawings

This appendix shows the actual design drawings that were used for an IFB along with the design specifications in Appendix D. They include a cover sheet with legend, floor plan with device locations, door schedule, two riser diagrams, and a panel layout detail. These are some of the types of drawings that can be submitted; others are also appropriate based on the project. Also, the layout and format are a matter of personal taste.

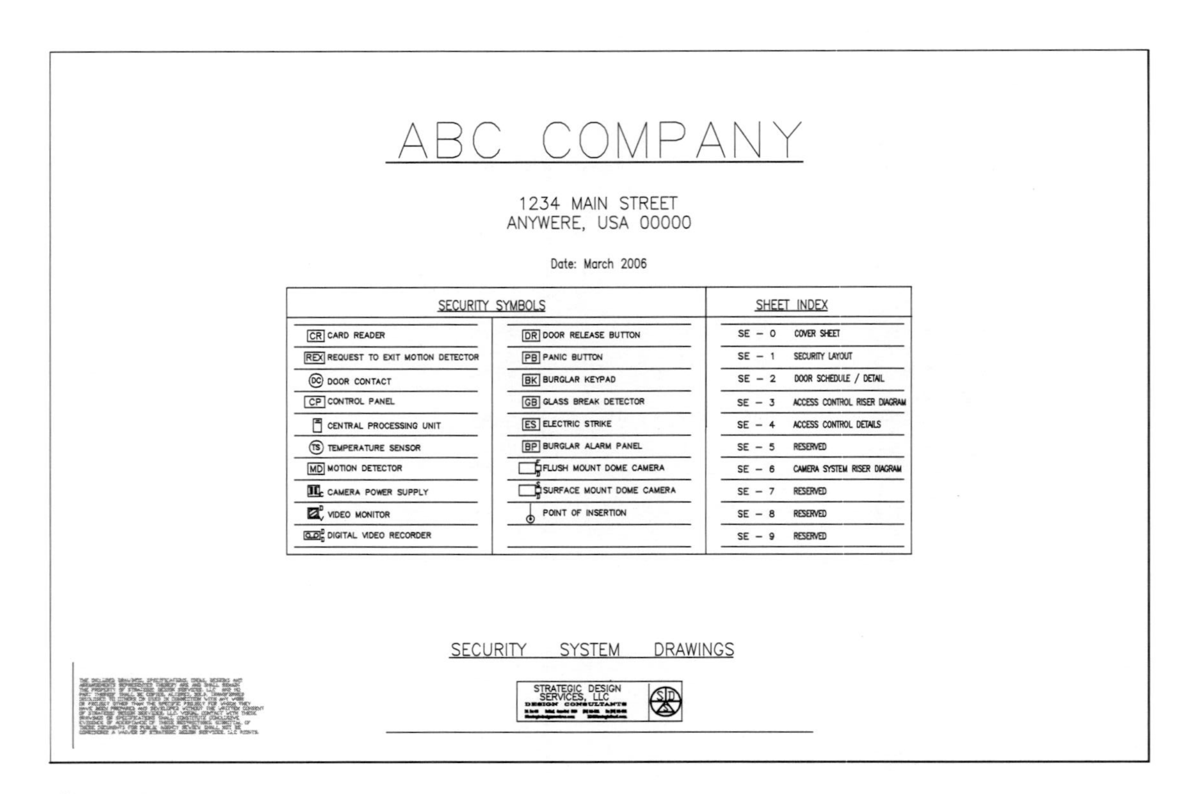

Figure E-1 Cover sheet with legend

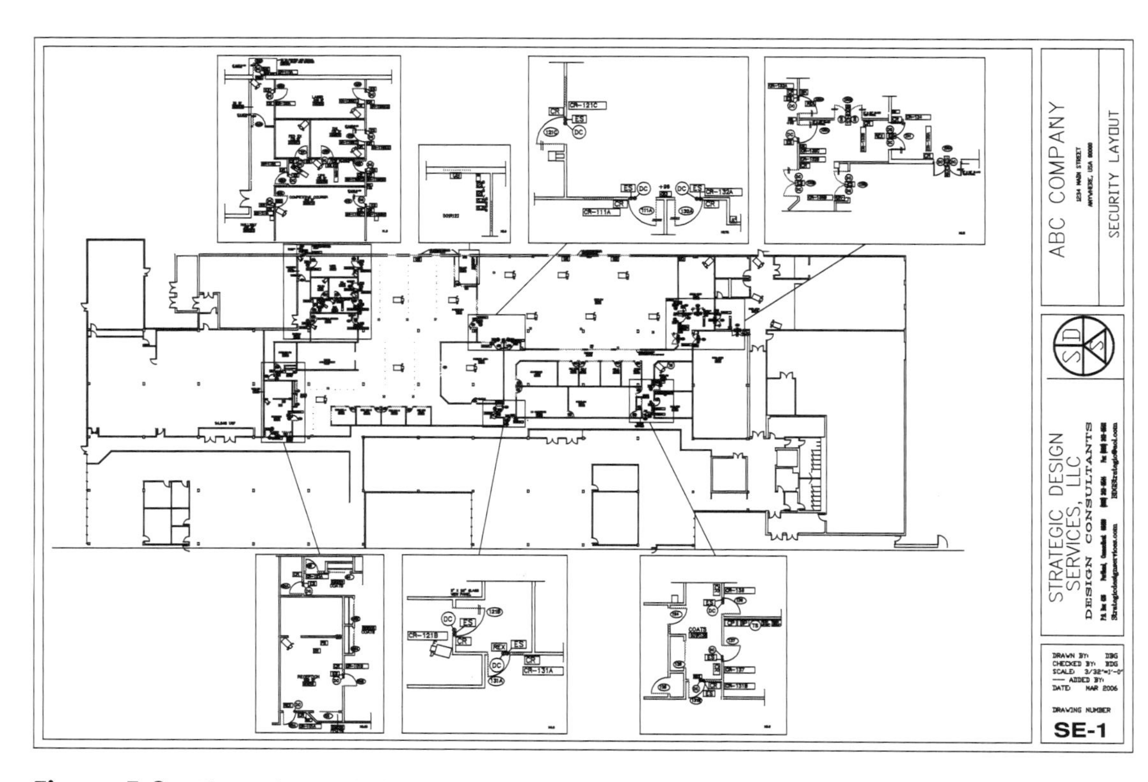

Figure E-2 Floor plan with device locations

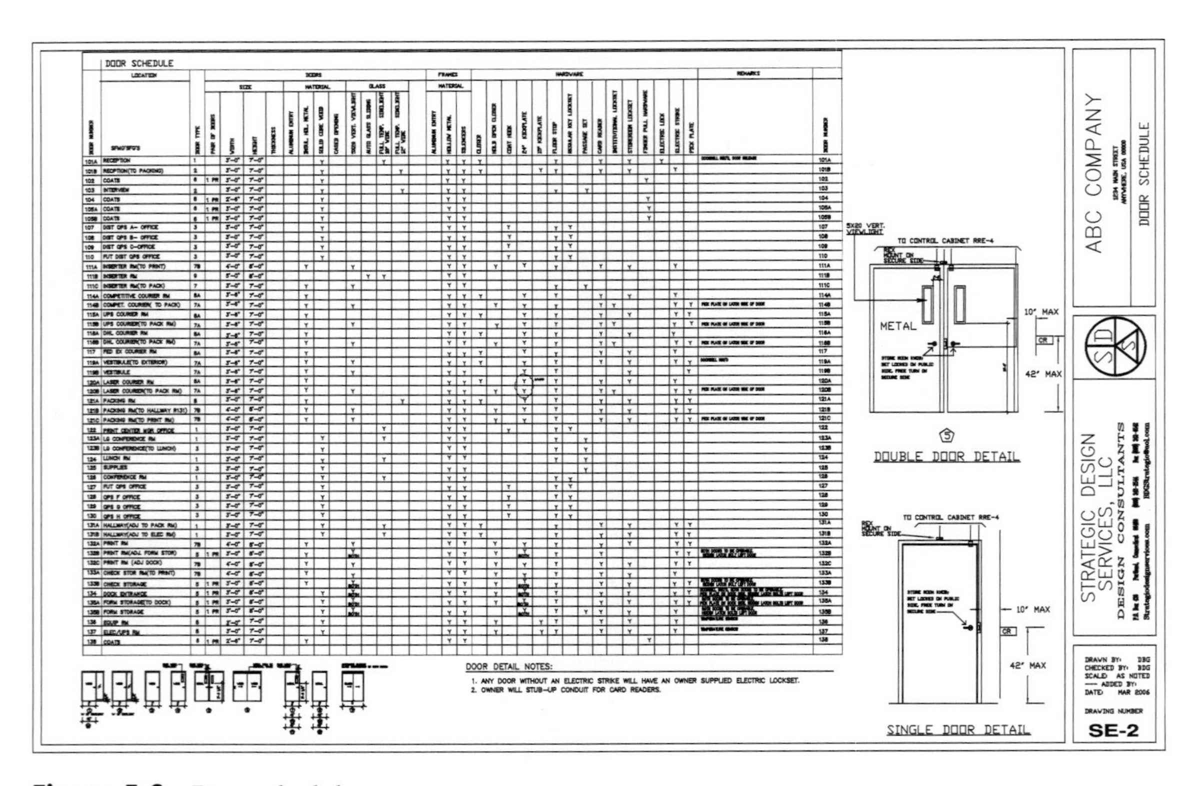

Figure E-3 Door schedule

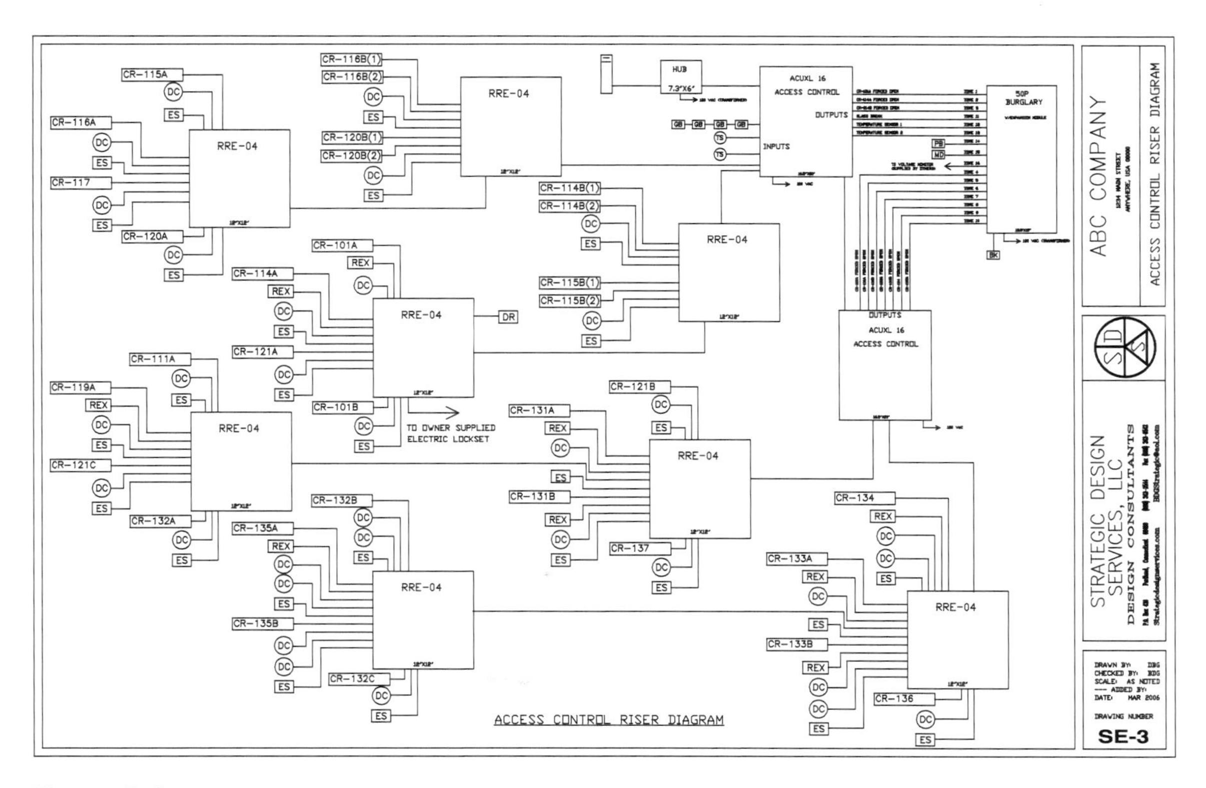

Figure E-4 Access control riser diagram

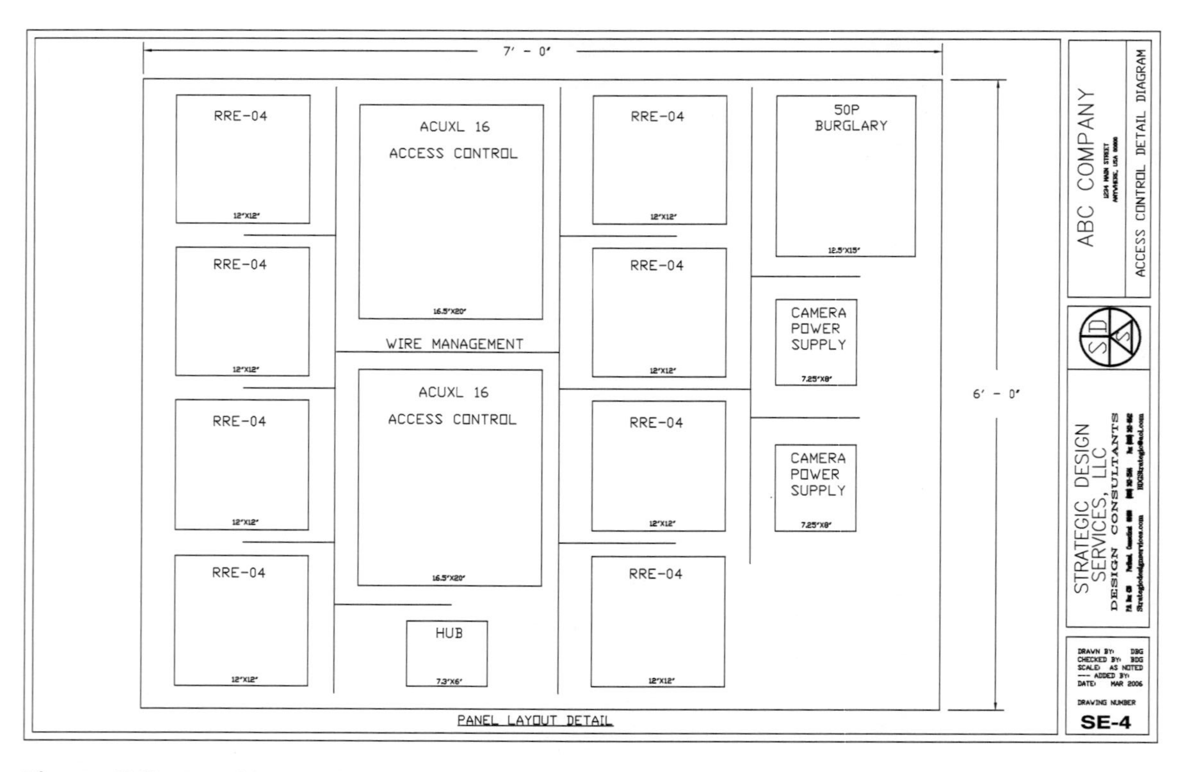

Figure E-5 Panel layout detail

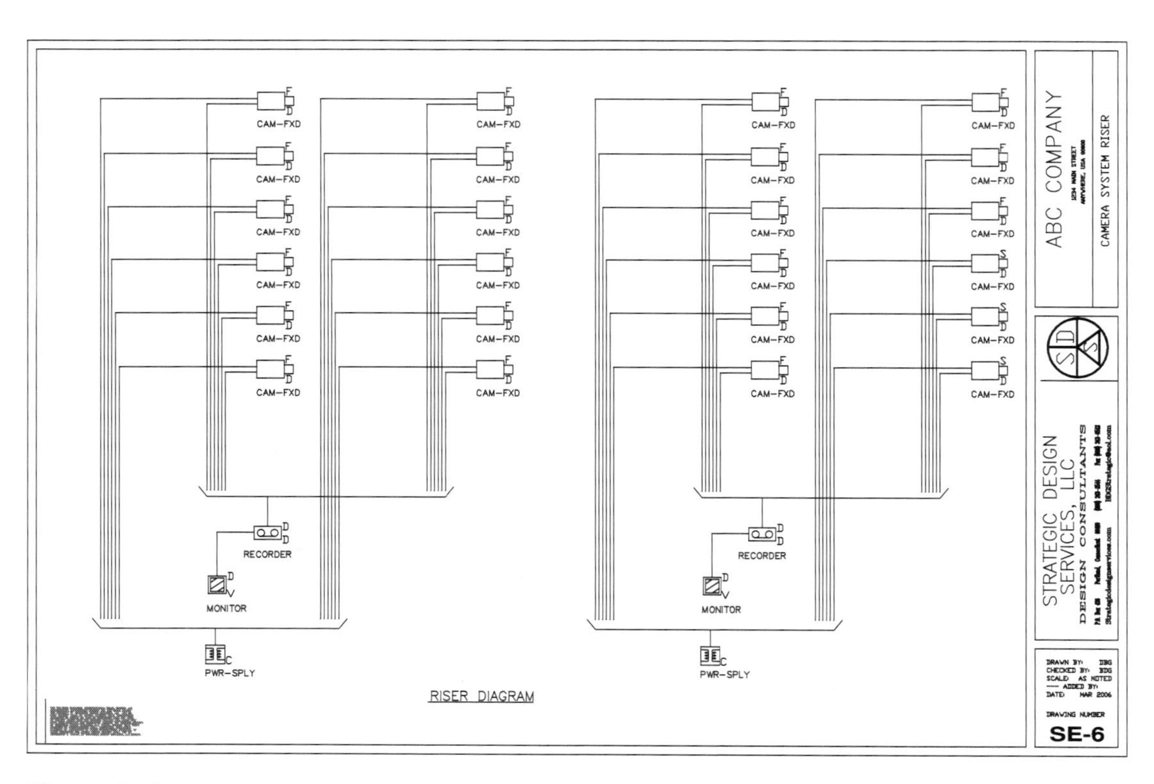

Figure E-6 CCTV riser diagram

Appendix F
Sample As-Built Drawings

This appendix shows the actual as-built drawings submitted by an integrator for the project designed using the specifications and drawings in Appendices D and E. They include a cover page, floor plan with cable runs, riser diagram, point-to-point wiring diagrams, and device wiring details. All of these drawings can be submitted as shop drawings before the project begins except the floor plan with cable runs and, in some cases, the device wiring details. Other types of drawings may also be appropriate to be submitted, depending on the project. The layout and format are also a matter of personal taste.

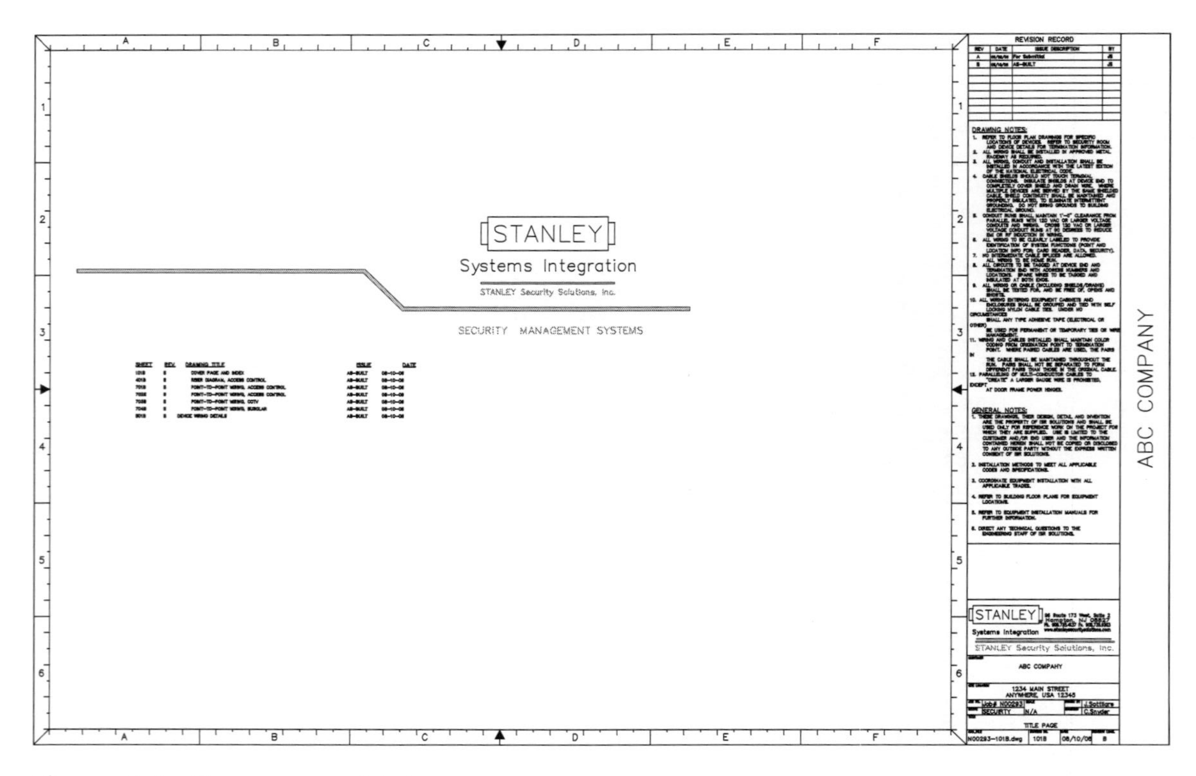

Figure F-1 Cover page and index

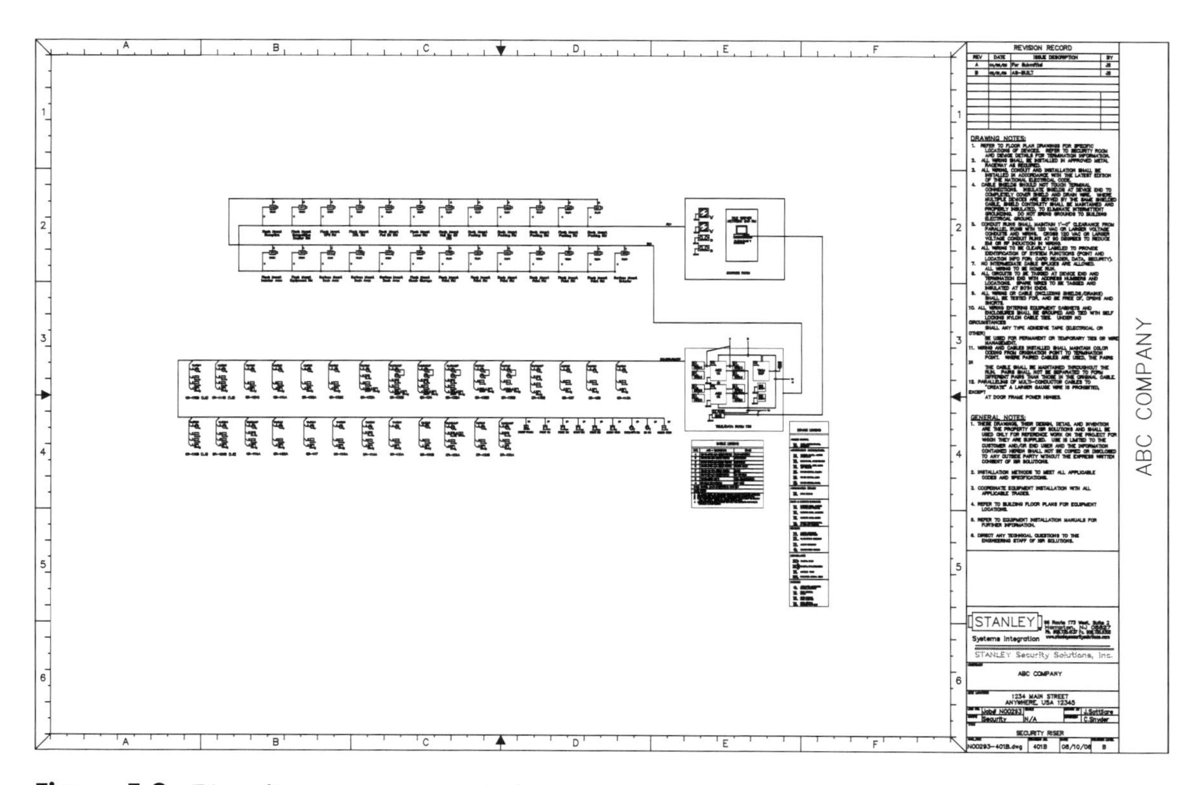

Figure F-2 Riser diagram, access control

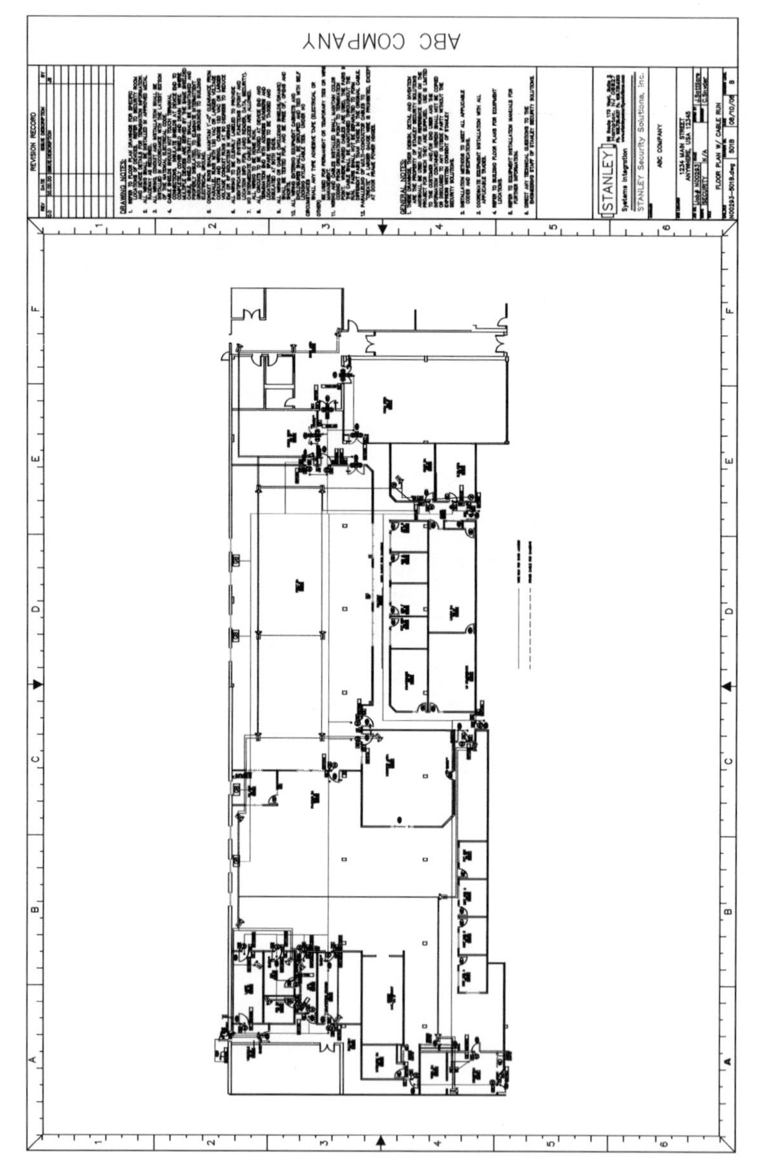

Figure F-3 Floor plan with cable runs

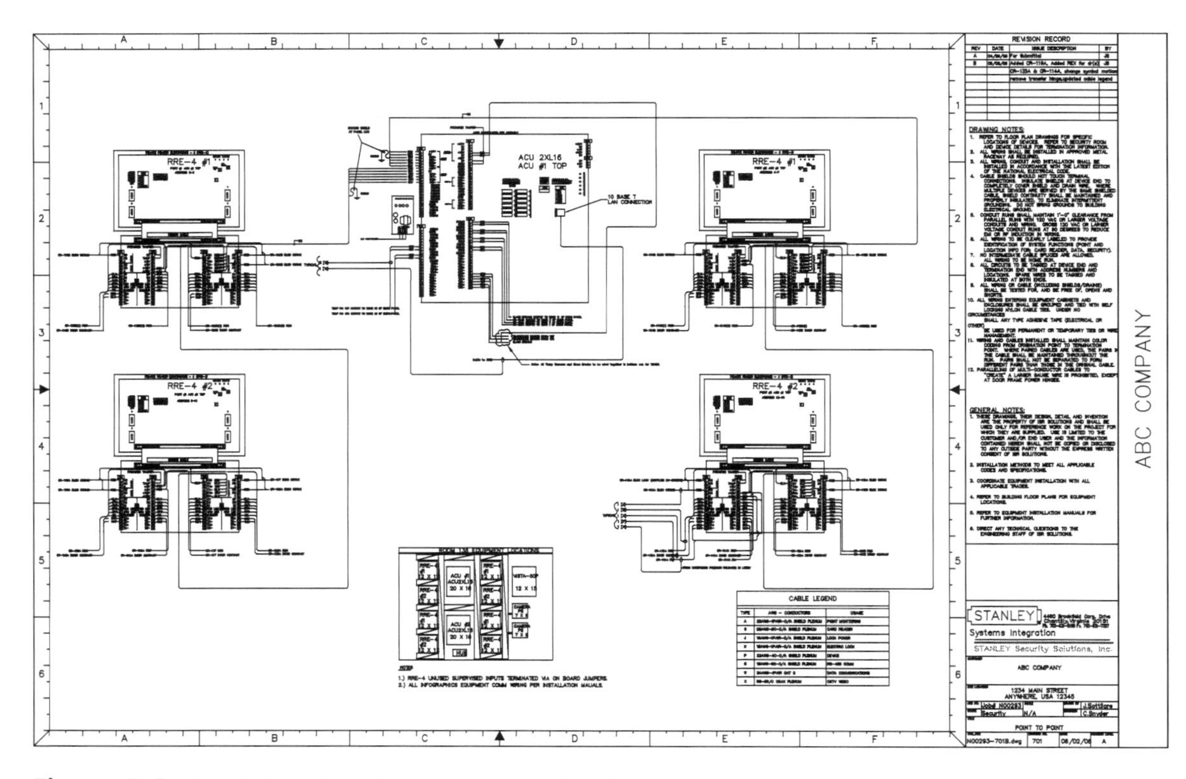

Figure F-4 Point-to-point wiring, access control 1

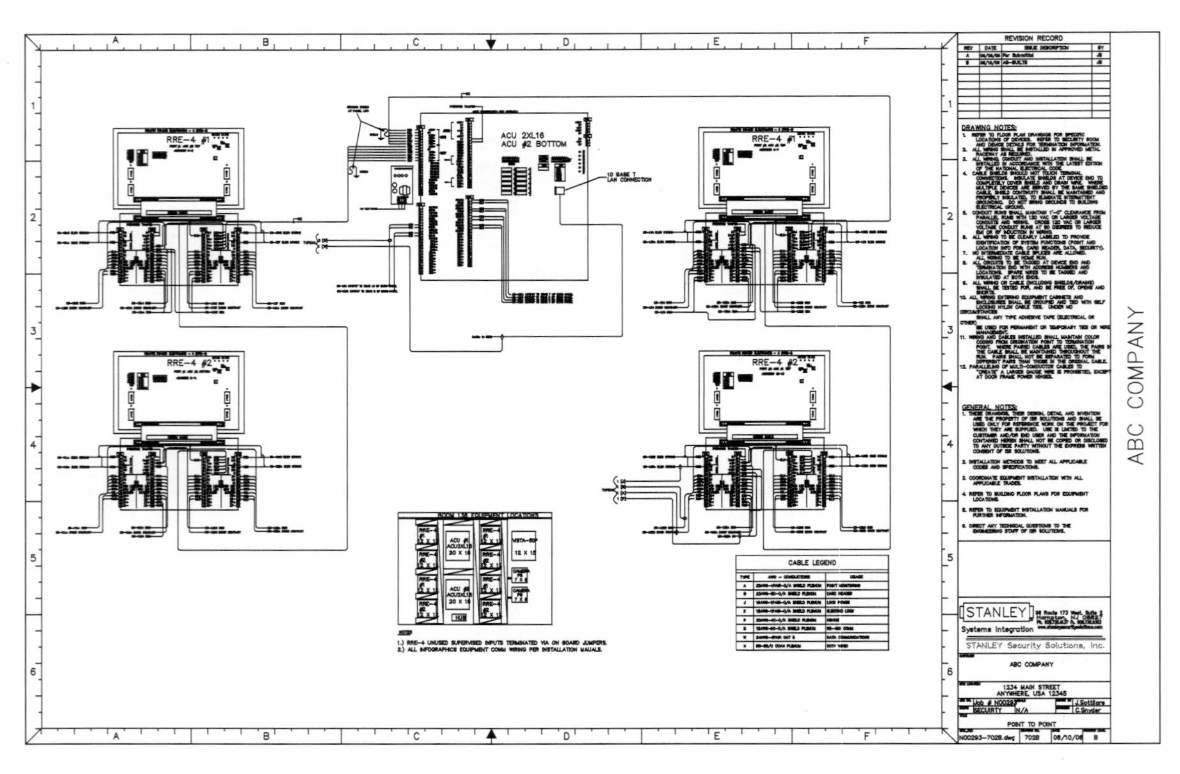

Figure F-5 Point-to-point wiring, access control 2

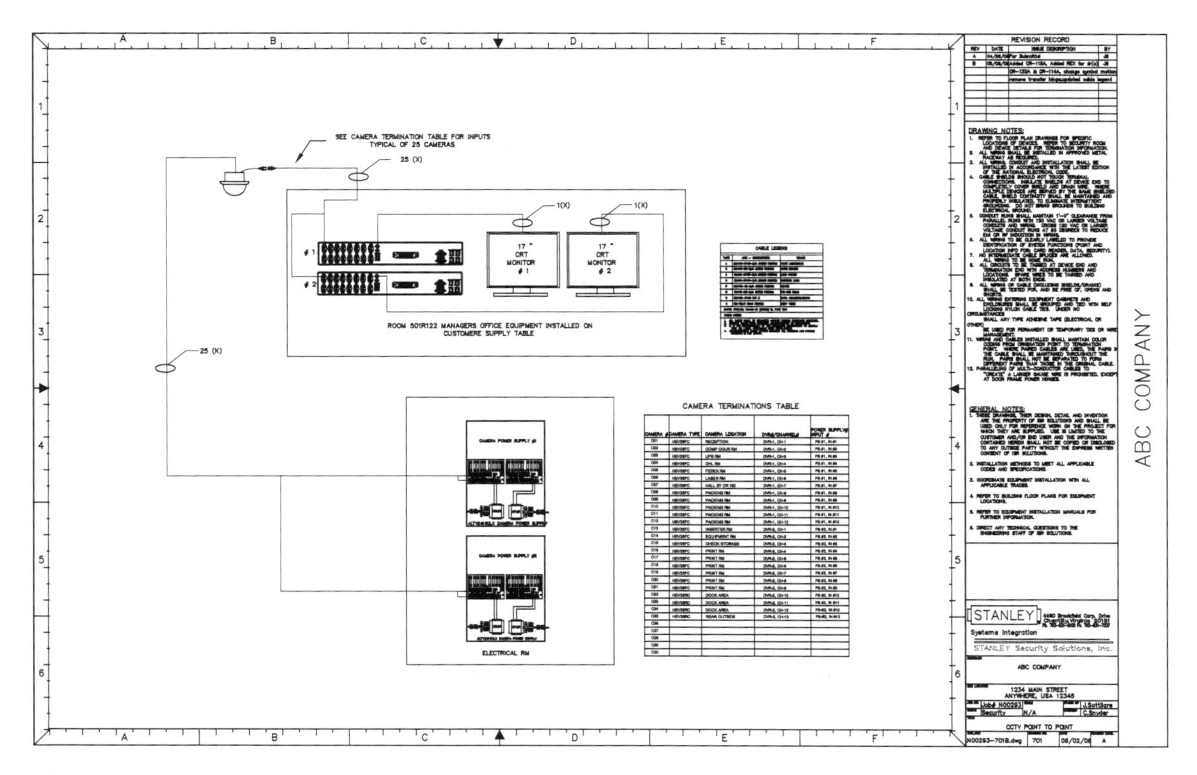

Figure F-6 Point-to-point wiring, CCTV

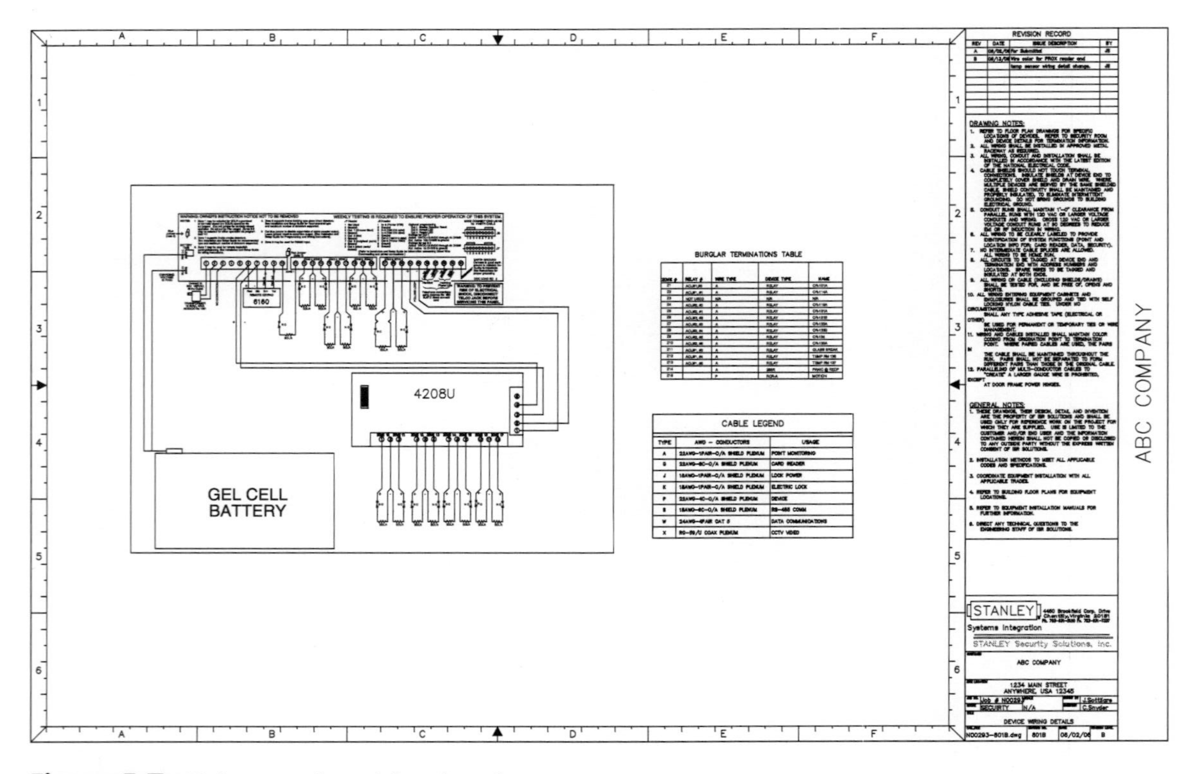

Figure F-7 Point-to-point wiring, burglary

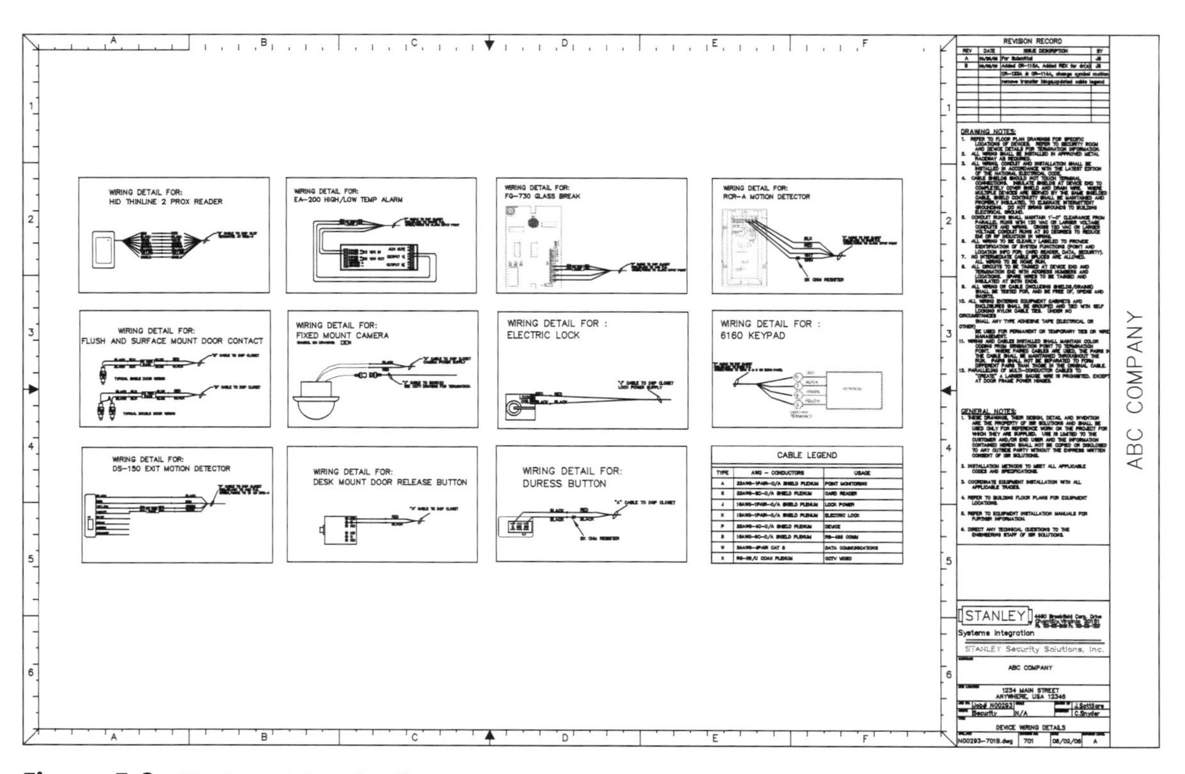

Figure F-8 Device wiring detail

Index

Printed and bound by CPI Group (UK) Ltd, Croydon, CR0 4YY
13/05/2025
01869611-0001